Critical Plant Studies in Taiwan

Critical Plant Studies

Series Editor

Douglas Vakoch

This series calls us to re-examine in fundamental ways our understanding of and engagement with plants, drawing on diverse disciplinary perspectives. The series encourages work grounded in the humanities and social sciences that provides innovative reformulations of the scope and practice of critical plant studies. Books in this series include both monographs and edited volumes that target academic audiences. To introduce critical plant studies to readers not familiar with this field, the series publishes work that is relevant to those engaged in critical plant studies, while also being of interest to scholars from the author's primary discipline. Among the books of special interest for the series are those that examine plants with reference to particular countries or regions of the world, or with respect to specific cultural, philosophical, religious, or literary traditions. Contemporary and historical works are equally appropriate. We especially welcome books that bridge academia and activism.

Recent Titles in the Series

Critical Plant Studies in Taiwan, edited by Iping Liang

Human-Plant Entanglement and Vegetal Agency in the Poetry of Thomas Hardy and Sylvia Plath, by Dilek Bulut Sarıkaya

Critical Plant Studies in Taiwan

Edited by

Iping Liang

LEXINGTON BOOKS
Lanham • Boulder • New York • London

Published by Lexington Books
An imprint of The Rowman & Littlefield Publishing Group, Inc.
4501 Forbes Boulevard, Suite 200, Lanham, Maryland 20706
www.rowman.com

86-90 Paul Street, London EC2A 4NE

British Library Cataloguing in Publication Information Available

Library of Congress Cataloging-in-Publication Data

Names: Liang, Iping Joy, 1961- editor.
Title: Critical plant studies in Taiwan / edited by Iping Liang.
Description: Lanham : Lexington Books, [2024] | Series: Critical plant studies | Includes bibliographical references and index.
Identifiers: LCCN 2023058951 (print) | LCCN 2023058952 (ebook) | ISBN 9781666935363 (cloth) | ISBN 9781666935370 (ebook)
Subjects: LCSH: Ecocriticism--Taiwan. | Plants--Taiwan--History. | Plants in literature. | Chinese literature--Taiwan--History and criticism. | BISAC: NATURE / Plants / Flowers | NATURE / Essays
Classification: LCC PN98.E36 C75 2024 (print) | LCC PN98.E36 (ebook) | DDC 809/.93364--dc23/eng/20240214
LC record available at https://lccn.loc.gov/2023058951
LC ebook record available at https://lccn.loc.gov/2023058952

Contents

Acknowledgments

I love plants. I grew up walking by banyan trees, camphor trees, hibiscus groves, and so on in southern Taiwan. As a child, I loved to see leaves fluttering against the blue skies. It was a most sensational moment in the military village where we lived. So, the first "thanks" should go to the trees—my "companion species." The idea of the collection was derived from the critical plant studies study group which I received government funds to operate between summer 2020 to summer 2023. It is needless to say that I am very grateful for the financial support. For three years, we met every other month to examine literary texts and their representations of plants. Five authors of the collection gave the first presentation of the individual chapter at the study group. I thank each of you—Weibon, Rose, Pei-Wen, Li-hsin, and CJ—thank you for coming along and presenting your work in the most detailed and engaging manner. I also thank Dr. Theodore Richard for his willingness to take on the novel, *Lord of Formosa*, by the Dutch writer, Joyce Bergvelt. His chapter sets Taiwanese sugarcane plantations at the forefront of plant mercantilism on the island, thus ushering in the following chapters on Formosa oolong tea trade and the war over the camphor. I also thank Yih-Ren, Pagung, Hsinya, and Cha-Hua for their agreement to have their co-authored journal essay included in this collection.

In addition, I thank Chia-ju's introduction to Prof. Steve Roddy, who then agreed to make a contribution on tea poetry to the volume. His chapter helps us realize the ecological degradation caused by monocrop tea plantations at the time. Finally, I'd like to give my gratitude to Li-Ru, Ya-feng, and Yalan—whose meticulous reading and profound insight enrich the book immensely. Last but not least, I also thank the Lexington editor, Mrs. Courtney Morales, for her patience, guidance, and trust: the book will not be concretized without her help and perseverance. Besides, I'd express my gratitude to the project assistant, Frank, whose help was most critical and timely. The final and deepest appreciation would go to my family—my parents, mother-in-law, husband, sisters, nieces, John and Emily—thank you; thank you. Your love and support keeps me going and this collection is dedicated to each of you.

Introduction

Iping Liang

In his seminal essay, "The Climate of History: Four Theses," the prominent postcolonial Indian historian Dipesh Chakrabarty argues that "the reality of global warming spells the collapse of the age-old humanist distinction between natural history and human history."[1] As Chakrabarty puts it, there is no separation between natural history and human history and this collection of essays aims to examine the intertwinement of the natural—the vegetal and the botanical to be specific—and the human in Taiwan from the perspective of critical plant studies.

Geographically situated between the Japanese Archipelago (Nihon retto) to the north and the Philippines to the south, Taiwan, as the mid-point of the Western Pacific Island Arc, is also known for vegetation diversity. According to environmental historian Jessica J. Lee,

> Taiwan is just eighty-nine miles wide, but in that distance it climbs nearly thirteen thousand feet from sea level. The jump to precipitous peaks creates a wealth of habitats, such that the island sustains a range of forests much vaster than its small footprint. The coasts are muffled with salt- and sun-soaked mangroves, and moving south, thick tropical jungle grows. The wet heat of a tropical rain forest thrums to temperate trees, and their hardwoods climb to pines. Boreal forest—with towering, size-of-a-house cathedral trees—grow up from the middle slopes of the island. Beyond the tree line, the mountains peter out to prairie, cane grasslands widening toward an alpine sky. Like topographical rings on a map, the trees array themselves by elevation.[2]

Due to its steep mountainous ranges, the island has a wide variety of botanical habitats—rising, by elevation, from the coastal mangroves, sugarcane plantations, tropical rain forests, temperate trees, broadleaf woodlands, to coniferous forests against "an alpine sky." Like South Africa and Madagascar, Taiwan is renowned for vegetation diversity. In *Flora of Taiwan*, it is noted that the island is home of approximately 4,000 species of vascular plants, 600 of which are ferns, 28 being gymnosperms, 2,400 being dicots, and 1,000 being monocots. According to Chun-che Huang and Guang-zhou Show, the island territory is only 0.0025 percent of the planet, but which provides 2.5 percent of global biodiversity. Due to its diverse geographies and varied climates, Taiwan would be considered as "the island of biodiversity."[3]

If the plant history of Taiwan signifies a crossroads of vegetation diversity, the human history is equally "crisscrossed." Being hailed as the "Ilha Formosa"—the beautiful island—by the Portuguese seamen sailing across the Taiwan Strait in 1545, Taiwan has gone through a process of "series colonization"[4] in terms of human history. Like the Philippines, Indonesia, Malaysia, and Singapore, Taiwan is one of the Southeast Asian Pacific islands that received waves of Western imperialism. In the wake of global history, Tonio Andrade places Taiwan in the context of the maritime trade network dominated by the Dutch East Indies Company (the VOC). In his book, *How Taiwan Became Chinese: Dutch, Spanish, and Han Colonization in the 17th Century* (2008), Andrade argues that the arrivals of Dutch, Spanish, and Han Chinese settlers—

> [S]hed light on one of the most important questions of global history: How do we understand the great colonial movements that have shaped our modern world? Historians have focused on European colonialism, paying little attention to non-western counter examples. I decided to study Taiwan because it is a place where European and non-European colonialism met, where two different civilizations encountered "people without history," and thus an ideal microcosm for understanding colonialism.[5]

Andrade thus shares the view with Chakrabarty: how the West and the East were connected by colonialism and capitalism. Taiwan is special in the way that the Dutch recruited Han Chinese settlers to migrate to Taiwan in the 1630s and introduced the system of "co-colonization"—"a Dutch military and administrative structure co-evolved with a much larger Chinese agricultural and commercial colony."[6] This paved the way for the cultivation of rice paddies and sugarcane plantations on the island. In addition, "wheat, ginger, tobacco, indigo, and many other crops could be grown on Taiwan, so that 'through the copious immigration of Chinese this place can in a few years be made into a small breadbox for the VOC's holdings in all the Indies.'"[7] The system of co-colonization literally introduced agriculture economy to Taiwan and had hence incorporated the island into the global trade network in the seventeenth century.

By crisscrossing the botanical and the global, this collection of essays investigates the "plant talks"—how plants make themselves seen, felt, cultivated, cut, collected, brewed, eaten, consumed, and traded—by different groups of people on the island. The multispecies "contact zone" between plants and humans in Taiwan manifests the "vegetal turn" in environmental studies, as proclaimed by both Joni Adamson and Cate Sandilands in the 2013 Association for the Study of Literature and Environment pre-conference seminar, "Vegetal Ecocriticism: The Question of 'The Plant.'" As they rightfully provoked, "recent works in cultural geography, science studies, object-oriented ontology, multispecies ethnography, philosophy, ethnobotany and other fields have raised especially acute questions about the ways in which plant agencies are included (or not) in social ecological imaginings of places, ethics, politics, and poetics."[8] Joining forces with the recent "mushrooming" of scholarship in critical plant studies,[9] this collection of essays will be the first of its kind that focuses on the "critical crossroads"[10] of Taiwan. Moreover, this critical intersectionality of Taiwan is intertwined in both botanical and human spheres. On the one hand, it is the critical "contact zone" that has connected the global history across China, Japan, Indonesia, the Philippines, Spain, and the Netherlands, as Tonio Andrade and other scholars, such as Anthony Reid and Leonard Blussé, have asserted. On the other hand, it is the only *islandic* biological refugium that connects the botanical history between the Himalayas and Taiwan east-westward, as well as that between Greenland and Taiwan north-southward. That is the reason why coniferous forests are found in Taiwan, because Taiwan had been the "island bridge" that once made the migration of alpine plants to the tropical zone possible in the last Ice Age. That is also the reason why there is a wealth of relict plants—the botanical survivors from the last Ice Age—on this tropical island.[11]

By intersecting the human with the vegetal, this collection of essays is divided into four sections to explore the entanglements between plants and humans in Taiwan from the seventeenth century and onward. The first section, "Western Explorations," examines in chronological order the Dutch, the British, and the American botanical explorations on the island. Due to its rich resources in rice, sugarcane, tea, and timber, Taiwan may have been imagined as another "spice island" from the perspective of "plant mercantilism"—the root cause that had connected "plants and empire" in the first place.[12] And this section examines the exploration and exploitation of natural resources, such as sugarcane, rice, tea, camphor, and so forth on the island.

Theodoor Richard's chapter, "Greedflation and Sugarcane Landscaping in Taiwan: A Transcultural Reading of *Lord of Formosa* by Joyce Bergvelt," details the "greedflation" of the VOC in relation to the cultivation of sugarcane plantations in the seventeenth century. Li-Ru Lu's chapter, "Empire's Nature: Economics, Ecology, and Formosan Tea in Nineteenth-Century Western Travelers' Natural Histories," presents in meticulous details the British and American imperial tea-prospecting in travel writings by Western explorers, diplomats, and naturalist who came to Taiwan in the 1860s. The second section, "Tea and Camphor Plantations," examines the representation of the tea and camphor plantations by local writers from Taiwan. Iping Liang's chapter, "Oriental Beauty: The Tea Plantations and Global Tea Trade in Nineteenth-Century Taiwan," examines the tea plantations and global tea trade in Jade Chen's novel, *The Merry Leaf*, from the perspective of the plantationocene. Steve Roddy's chapter, "From Resonance to Solastalgia: Tea Poems of Taiwan, ca. 1820–1920," reflects on the ecological degradation as a result of monocrop tea plantations in selected tea poems during the turn of the last century. Ya-feng Wu's chapter, "Brothers Locked in Strife: A Molecular Reading of *Blood Camphor*," gives a riveting account of the brutal fights between the Japanese and the Indigenous people over the extraction of camphor in Ying-Ming Chang's novel, *Blood Camphor*. By analyzing literary texts from local Taiwanese writers, this section investigates how local Taiwanese poets and novelists react to the changes in the environments and to the encroachments of the colonial regime.

The third section, "Afforestation and Conservation," focuses on the themes of military afforestation in Kinmen during the Cold War era (1949–1989), as well as the environmental conservation movement in the 1980s, highlighting the importance of casuarina and mangrove forests. Weibon Wu's chapter, "Beyond Cold War Afforestation: Trans-scalar Imaginary of Casuarinas," provides an insightful reading of both the history of casuarina afforestation in Kinmen and the close bonding between Kinmen residents and the casuarina forests. Rose Hsiu-li Juan's chapter, "Mangrove Taiwan: A Birdman's View of an Island Environmental Identity," reads closely Ke-Xiang Liu's mangrove poems in relation to the environment conservation movement in the 1980s.

The fourth section, "Posthuman Vegetalities," turns to alternative plantscapes in Taiwan since the 1990s. Pei-Wen Clio Kao's chapter, "Politics of Femininity, Politics of Plants: The Roof Garden in Zhu Tianwen's "Fin de Siècle Splendor," turns to vegetal ecofeminism by examining the politics of rooftop gardens toward the end of the 1990s. Li-hsin Hsu's chapter, "'My Head Like a Mushroom Below": Chen Kehua's Vegetal Gothic and the Anthropocene," observes the ecogothic scenery of decay and decadence in Chen Kehua's selected poems. Kathryn Yalan Chang's chapter, "Reclaiming the Commons: An Ambient Poetics in Jessica J. Lee's *Two Trees Make a Forest,*" focuses on the vegetal common that is rendered possible by way of the ambient poetics in Lee's nature writing. Yih-Ren Lin et al.'s chapter, "The Cultural Narrative of Tayal's Millet Ark," gives an account of the oral tradition of the process of millet restoration in the Ataya tribe in northern Taiwan.

Finally, Chingshun Sheu's chapter, "The Call of the Hyperobjects: Plants as Zones of Aesthetic Causality in Tao Lin's *Taipei*," renders the seeming absence of plants in urban Taipei in the context of the underground mycorrhizal network and thus turns absence into presence. The five chapters—the ecofeminist rooftop garden, the ecogothic mushroom, the ambient eco-therapic narrative in the forest, the Indigenous millet restoration, and the underground mycorrhizal networks in urban Taipei—are able to render the postmodern and posthuman plantscapes in contemporary Taiwan.

Taken together as a whole, the twelve chapters in this collection would be able to provide both a macrocosm and a microcosm of the plantscapes in Taiwan—from the multiple angles of Western explorations, colonial bio-prospecting, global tea trade, Indigenous millet farming, and posthuman vegetalities that are related to the ecofeminist, the ecogothic, the eco-therapic, and the underground mycorrhizal network. In terms of specific plant species, it assembles the diverse "plant talks" in the materialist and metaphorical plantscapes of sugarcane farming, tea and camphor plantations, Indigenous Ataya millet fields, military casuarina forests, tropical mangrove forests, and the underground "wood wide web." Like the "topographical rings on a map," the literary and materialist "plant talks" array themselves layer by layer in the constellation of these essays—the first of its kind for critical plant studies in Taiwan.

NOTES

1. P. 201.
2. Pp. 5–6, *Two Trees Make a Forest: Travels among Taiwan's Mountains and Coasts in Search of My Family's Past*, 2020.
3. Pp. 3–4, *Flora of Taiwan*, Huang ed., 2004.
4. Please see Shih Shu-mei, "Theory in a Relational World," p. 726.
5. *How Taiwan Became Chinese*, p. 2.
6. Ibid., p. 118.
7. Ibid., p. 122.
8. Please see "Vegetal Ecocriticism: The Question of the Plant."
9. Please see among others, Michael Marder, *Plant-Thinking: A Philosophy of Plant Life*, Columbia University Press, 2013; Randy Laist, editor, *Plants and Literature: Essays in Critical Plant Studies*, Rodopi, 2013; John. Ryan, Monica Gagliano, and Patricia Vieira, editors, *The Green Thread: Dialogues with the Vegetal World*, Lexington, 2016; John Ryan, *Posthuman Plants: Rethinking the Vegetal through Culture, Art, and Poetry,* Common Ground Publishing, 2015; Monica Gagliano, John Ryan, and Patricia Vieira, editors, *The Language of Plants: Science, Philosophy, Literature*, University of Minnesota Press, 2017; Suzanne Simard, *Finding the Mother Tree*: *Discovering the Wisdom of the Forest*, Vintage, 2021.
10. This is the title of Anthony Reid's book, *A History of Southeast Asia: Critical Crossroads*, Blackwell, 2015.
11. For details, please see Zhi-Chien Yu, *Worldviews: The Origin and Journey of the Montane Plants in Taiwan,* Spring Mountain, 2020.

12. Please see Londa Schiebinger, "Plants and Empire: Colonial Bioprospecting in the Atlantic World," Harvard University Press, 2004.

BIBLIOGRAPHY

Adamson, Joni and Cate Sandilands, "Vegetal Ecocriticism: The Question of the Plant." The seminar given in the 2013 Association for the Study of Literature and Environment." https://asle.ku.edu/Preconference/pdf/Vegetal_Ecocriticism_The_Question_of_ThePlant.pdf

Andrade, Tonio. *How Taiwan Became Chinese: Dutch, Spanish, and Han Colonization in the Seventeenth Century*. Columbia University Press, 2008.

Chakrabarty, Dipesh. "The Climate of History: Four Theses," *Critical Inquiry* vol. 35, no. 2 (Winter 2009), pp. 197–222.

Huang, Tseng-chieng, editor, *Flora of Taiwan*, 1984, 2nd edition, 2004.

Lee, Jessica J. *Two Trees Make a Forest: Travels among Taiwan's Mountains and Coasts in Search of My Family's Past*, Catapult, 2020.

Reid, Anthony. *A History of Southeast Asia: Critical Crossroads*, Blackwell, 2015.

Schiebinger, Londa, *Plants and Empire: Colonial Bioprospecting in the Atlantic World,* Harvard University Press, 2004.

Shih, Shu-Mei, "Theory in a Relational World," *Comparative Literature Studies,* vol. 53, no. 4, Special Issue: The Indiscipline of Comparison, 2016, pp. 722–746.

Chapter 1

Greedflation and Sugarcane Landscaping in Taiwan

A Transcultural Reading of Lord of Formosa by Joyce Bergvelt

Theodoor A. M. Richard

There is a re-occurring advertisement for dental work that from time to time beautifies the sides of city buses in the municipality of Taichung, Taiwan. The advertisement has been the same over the years so I assume that it has been effective in drawing in customers. It is of a very simple design and largely shows an elderly man in his late seventies, plainly dressed in an outfit of long sleeve shirt and pants very common for elderly men in Taiwan. This old man is looking perky and widely grinning at us viewers, as he is about to set his bright white teeth into a stick of shaved sugarcane that he is holding up in his right hand, hovering closely to his face. Clearly the image communicates an indelibly attractive promise of the pleasure of being able to crunch the hard fibery sugarcane and reaping the sweet reward of tasting its delicious juice as the result of the biting, chewing and savoring allowed by the strong and well-fitting dentals that the advertiser can provide for you, regardless of age. It is a commonly known fact in Taiwan that crunching the sugarcane stem requires strong teeth so the promise of artificial dentals that are able to give you that excitement without destabilizing, does raise public expectation.

As the bus pulled away from the stoplight that it had waited for, I thought again of the news that I had read in the digital version of *Fortune*. The news was about a global strategist of the reputed French bank Société Générale, named Albert Edwards, who had written a pamphlet in the bank's *Global Strategy Weekly* of April 2023, warning against a phenomenon called

Greedflation—the creation of inflation by corporations that use the rise of raw material costs during a crisis to raise prices and expand profit margins exorbitantly. According to *Fortune*, Edwards wrote: "After four decades of working in finance, he's never seen anything like the 'unprecedented' and 'astonishing' levels of corporate *Greedflation* in this economic cycle."[1] There is no doubt that inflation levels have skyrocketed worldwide, as every basic need has become so much more expensive, from energy and fuel prices at the pump to food supplies in the supermarket. Central banks have tried to counteract this inflationary drive macro-economically by making borrowing money more expensive, but this has obviously not worked and only driven up prices and economic distress even further. The simple reason why this traditional macro-economic measure has not worked is because the cause of the inflation is micro-economic: the decision by market dominant corporations to increase their prices beyond a reasonable cost-income balance, just because they can. As Weber and Wasner explain why large firms can hike prices in an emergency: "Firms only raise prices when they are confident that their market shares will not be harmed."[2] The CEO of BP, Mr. Bernard Looney, commented on its boosted profits of GBP 2,4 billion in Q3 of 2021 in this way: "When the market is strong, when oil prices are strong and when gas prices are strong, this is literally a cash machine."[3] Oil companies would go on to double their profits over 2022, with earnings by BP of USD 28 billion, by Shell of USD 40 billion, and by Exxon-Mobil of USD 56 billion, up by 143 percent from 2021.[4] And it is not just oil companies, but banks as well and even supermarket chains. All this has led the editorial board of *The Guardian* to declare that corporate greed is indeed causing inflation.[5] The data published by the US *Bureau of Economic Affairs* (BEA) on industry-wide whole economy corporate profits in the US show that they increased over 2021 by a whopping 22.6 percent and more modestly, but still quite up, over 2022 by 6.6 percent.[6] Albert Edwards used the information provided by the BEA in his above pamphlet in a graph showing that even though labor cost rose very little, and non-labor cost in fact dropped substantially in 2022, the whole economy profit margins in the US increased dramatically. So the general narrative that the media brought in 2022, saying that the Ukraine war was the cause for shortage of fuels and foods, which in turn would force us to pay much more for everything, is simply false. We have been suffering grossly from the effects of corporate greed instead. And why? Just because the corporations thought they could increase prices in their relentless pursuit of profits during a world crisis, and this irrespective of the consequences this might have for the global social, cultural, and ecological landscape that they saw as their sourcing- and/or selling market.

What then does this exposé on *Greedflation* have to do with sugarcane in Taiwan? It is exactly this juncture that will lie at the core of the argumentation in this chapter. Here I will argue that from an environmental historical perspective, the sugarcane landscaping in Taiwan as it originated in the seventeenth century has everything to do with corporate greed, at least as an early case thereof, and avant-la-lettre. In contemporary Taiwan, sugarcane cultivation is considered as a traditional agricultural business and its products of brown and refined sugar have been part of Taiwanese products. Every traditional marketplace has at least one stand that sells either sugarcane juice that when cooled or frozen, is considered the perfect antidote to the hot summer in Taiwan, or the shaven stems of the raw sugarcane for chewing and savoring as a sweet delight. Taiwan Sugar Corporation (TSC) was established in 1946 to unite all sugar industries under the nationalist rule. Indeed, the sugarcane business may have flourished for around 400 years in Taiwan, and as such has become a classic mind-picture of old-style rurality on the island. But very few Taiwanese may know that it was in fact a foreign practice imported into Taiwan by the corporation VOC (*Verenigde Oostindische Compagnie*) in the early seventeenth century to boost the profits from its trade in the region. I will set out to show that the conditions established around the business of sugarcane and in its wake, rice-cultivation upon its introduction in Taiwan, has dramatically changed the landscaping in Taiwan—ecologically, socially, and culturally. I will also argue that this phenomenon has arisen as a result of trade policies by the VOC and its competitors to power in Taiwan, the Zheng family of whom the man known as Coxinga had the most influence in this process. These trade policies have included practices and acts of piracy and war, on both sides of the Taiwan Strait, which at that time seemed to be commonly regarded as normal activities comprised in the whole of trading endeavors. As such, the sugarcane landscaping in Taiwan as re-imagined from an environmental historical perspective, becomes the transcultural outflow of trade driven by unchecked corporate greed. And I will explore this issue through close reading of the historical novel *Lord of Formosa* by Joyce Bergvelt.

Born in the Netherlands, Bergvelt followed her family to Japan at the age of 10 and later to England and Taiwan, where she studied Chinese for one year and a half. She then went to England to obtain a degree in Chinese Studies from the University of Durham. Later on she studied Journalism in Sydney and worked freelance in Dubai for a Dutch magazine. The debut novel *Lord of Formosa* was published in Dutch in 2015 under the title *Formosa, voorgoed verloren.* The Dutch version was longlisted for the Hebban Debut prize in 2016. The English version was published by Camphor Press in 2018 and was the runner-up for Asian Books Blog Book of the Year in 2019. A Traditional Chinese language edition was published by Avanguard/

Qianwei Publishing House in 2023. Bergvelt has since published a second novel in the Dutch language in 2022.[7] *Lord of Formosa* has been positively reviewed by Hilton Yip for the *Asian Review of Books* and Bradley Winterton for the *Taipei Times* (2018). Yip calls it "a sweeping, fast-paced novel" and "a compelling account of the nobleman warrior and pirate Koxinga." According to Yip, "the narrative moves smoothly" and "readers get a good sense of Taiwan as a frontier settlement." Yip regards it as a "fantastic action-packed novel" and commends Bergvelt "for a fine effort in bringing to the fore both a restored Koxinga," as well as for highlighting the role of the VOC in the development of Taiwan. Winterton stresses that "this very fine novel is as awash with as much researched detail as it is with vividly imagined scenes" and claims that this "excellent new edition of the English version" truly delivers an "outstanding reading."

THE HISTORICAL CONTEXT: THE VOC, CONXINGA, AND GREED

In sixteenth-century Europe, the political landscape was being shaped by clashing interests of religion and taxes. The small northern European territory that was known as the 19 Provinces of the Netherlands had been conquered by the Spanish who were Roman Catholic in religion. As the protestant religion however rapidly gained popularity in the 19 provinces, especially the most northern ones, religious intolerance abounded. This movement seemed to coincide with increased dissatisfaction of the nobility in that territory with the levy of heavy taxes by the Spanish court. The discontent grew to such a level that the 19 provinces revolted against the Spanish rule in 1568 which ensued the 80 Year War. The revolt resulted in the written abjuration by several of these provinces of their oath of loyalty to the Spanish king by the execution of the *Plakkaat van Verlatinghe* (the Act of Abjuration), declared on July 26, 1581. Consequently, seven of these provinces united by adopting sovereignty away from the king and upon and over themselves by the establishment of a new republic entity called the *Republiek der Zeven Verenigde Nederlanden* (Republic of the Seven United Netherlands, or: The Republic) in 1588. All this warring was detrimental to the imports into the northern Netherlands of precious spices from the East, especially pepper, the trade of which had been monopolized by the Portuguese and Spanish. In order to circumvent this trade embargo and import the spices directly, the Republic decided to venture out and send their own trading expeditions to East Asia. Several trading companies were established and did manage to establish such trade, but they started to compete amongst themselves, bringing down profits and the levies that could be paid to the Republic. This was of grave concern to the governing body of the Republic, *de Staten Generaal* (SG), and especially its *Raadpensionaris* at the time, Johan van Oldenbarnevelt, who was second executive in charge. Van Oldenbarnevelt felt that in this way, not enough funds were being raised to finance the ongoing war with the Spanish and he forced these companies to merge into one company which was to become the VOC on March 20, 1602.[8]

The VOC was one of the two first international examples of a limited liability company. Although it was established by decision of the SG, it was not owned by the government. Its capital was raised exclusively among private shareholders who were free citizen businessmen from the under-chambers of the SG, with the chamber of Amsterdam putting in the most money to an amount of more than 3.5 million guilders which was also more than half of the total share capital of 6.4 million guilders; the free tradability of the VOC shares also resulted in the birth of the world's first stock exchange in Amsterdam.[9] These companies spearheaded the development of the currently characteristic elements of modern corporations: "a permanent capital, legal personhood, separation of ownership and management, limited liability for shareholders and for directors, and tradable shares."[10] Moreover, the VOC is generally considered the world's first multinational corporation because it had several establishments in East Asia, most notably in Indonesia, Formosa, Japan, and Ceylon (currently Sri Lanka) and can be considered as the richest and most powerful company at the time. Apart from its core activity of trading, buying, transporting, and selling goods, "by the late 17th Century [the VOC was involved in] construction, sugar-refining, cloth manufacturing, tobacco curing, weaving, glass making, distilling, [and] brewing."[11] And as such, it pioneered another feature of modern multinational corporations: the unconditional drive for profitability that when unchecked, steers us into the realm of corporate greed, or *Greedflation*, as it was expressed in the words of Shell's CEO Wael Sawan when discussing the corporate strategy on switching away from fossil fuels into renewables: "We cannot justify going for a low return. Absolutely, we want to continue to go for lower and lower and lower carbon, but it has to be profitable."[12] In the same way, the VOC decided on all its matters of business by the yardstick of profitability. Pieter de la Court, a contemporaneous critic of the VOC expressed it in the 1669 extended edition of his pamphlet *Interest van Holland* as follows: "The true *Interest* [of the VOC] consists in seeking the most benefit for its common Participants . . . and, in short, in making the *most* profits with the *least* trade and navigation,"[13] thus neglecting the interest of the common good of the Republic.

To further understand the complicated yet pivotal role that the VOC had played in the history of sugarcane landscaping in Taiwan, it needs to be explained that the VOC was not an ordinary corporation as we may know it in the modern sense of the word. As already mentioned, the VOC was established with the very particular purpose to develop independent trade and imports of in first instance, mostly spices from East Asia, and maximize profits while doing so, in order to bring optimum levies to the Republic that were needed to finance the ongoing war against Spain. To facilitate this goal, the Republic's SG gave the VOC an explicit written monopoly on such trade and a clear *octroi*, or authorization,[14] to use any means necessary, including conquering lands, signing treaties with foreign powers, and even waging war, especially against Spain and Portugal.[15] The VOC, loaded with fresh capital from its shareholders, zealously set out to do so and managed already in the early years after its incorporation to wipe out much of the Portuguese trade by hindering Portuguese ships from going in or out of their then-trade colony Macau and bringing their trade with Manila "to almost a complete halt,"[16] therewith securing that there would be no more ongoing exports by the Portuguese to Europe. From there the VOC went on to develop many trade posts, fortresses and even local headquarters all over East Asia, including Formosa, and became such a huge conglomerate of trade, territorial, and political interests in the course of the seventeenth century, which Arthur Weststeijn regarded as a "Company-State," a "body politics that was neither corporation nor empire."[17] The VOC can in fact be argued to be the world's first example of a Private Public Partnership (PPP). As such, the VOC was basically empowered to do whatever it wanted to enrich itself without competition from free trade or the burden of colonization, for which it was also criticized even at its own time, but to no avail. In 1650, the VOC directors (or *Heeren XVII*) issued General Instructions to all its employees in its local headquarters in Batavia, Indonesia, that specifically made this clear: "The entire welfare of the East India Company [depends on the Company being able to] enjoy privately, to the exclusion of all others, the fruits of the trade conferred to her alone."[18]

Competing European maritime powers were however not the only obstacle for the VOC. The VOC had to contend with the fact that the waters now forming the Taiwan Strait and the East- and South China seas, were ruled by Chinese pirate bands, of which the Zheng family clan was by far the mightiest. The Zheng family merchant pirate estate was started by Zheng Zhilong (1604–1661) who developed himself from a merchant apprentice into the unopposed leader of the Chinese pirates in the region,[19] whose "trading empire and personal fleet extended all the way to present-day Singapore."[20] Piracy, or "sea raiding" as it is elegantly called by Robert Antony, was considered by many Southeast Asians as "a way of life closely tied to war, slavery, and trade. . . . While viewed as a criminal act by Europeans, raiding was actually a respectable profession pursued not only by individuals but by entire communities and even kingdoms. It was a common means for warriors and chiefs to increase their power and prestige."[21] This explains why Zheng Zhilong was able to move so smoothly from being a merchant pirate into an official Ming Dynasty general in 1628,[22] even though this did not stop him or his son, Zheng Chenggong, or, Coxinga (which is the name I will refer to henceforth,) from continuing to build their trade empire. Not only did they raid all ships that tried to do business in their maritime turf, they also sold free passes to fishermen and merchants while violently enforcing any sailing through without them.[23] Based on the historical analysis by Cheng Wei-Chung in his book, *War, Trade and Piracy in the China Seas, 1622–1683*, Richard Blakemore calls this "a fluid situation where aggression and diplomacy, trade and violence, could be and often were simultaneous activities. For the merchant warlords, this seems to have been cyclic: trade created their opportunities and their wealth, which made them powerful, and with that power they were able to exert more control over trade."[24]

This historical maritime tug of war that happened in the early to mid-seventeenth century between the VOC and the Zheng clan played out against the background of the civil warring in China between the Ming Dynasty and the Manchus. It is generally argued that Zheng Zhilong to some extent, but especially his eldest son Coxinga, were strong Ming Dynasty loyalists bestowed with grand military titles by the Ming Emperor, and that they put their military might in service of the fight against the invading Manchu armies which despite their efforts, would later conquer China and form the Qing Dynasty. However, as autobiographical information about the Zhengs is lacking, it remains to be contested what their real motives were. Indeed, historically they did fight on the side of the Ming Dynasty, although the elder Zheng decided to switch sides in 1646, much to the apparent dismay of his son, who continued to fight against the Manchus until his death in 1662. Pomeranz and Topik even call it "an overriding obsession" of Coxinga to restore the Ming Dynasty despite the fact that "most of his staggering wealth was located offshore."[25] But as already mentioned, sea-raiding and warring was not necessarily a matter of principle or loyalty to any one side for the pirates. These were merely further practical ways of increasing power and wealth. The Ming Emperors were faithful and notorious landlubbers who declared a Maritime Prohibition Act and sought the expansion of the Empire only over land so they left the seas open for pirates to profit.[26] Hence the strong prevalence of buccaneers in the area and the clear horizons for the Zhengs to accumulate such "staggering wealth." Thus considered, it seems fairly out of place to assume that notorious pirates as the Zhengs would suddenly become such pious imperial loyalists, especially where Zheng the father had earlier used his great fleet of war junks to defeat whatever navy the Ming Dynasty could muster in 1628. It can be argued that the father's acceptance of an appointment as Ming Imperial Admiral following such victory may also be read as an act of sheer opportunism that allowed the Zheng clan to expand its power even more in service of furthering the clannish wealth accumulation. This trait of opportunism seems to characterize him well considering that he changed sides to the Manchus later. How much this was all a ruse or strategy to secure and further expand the family power and wealth may remain a matter of speculation. And as such, it is one of the themes that Bergvelt readily uses to shape her story in *Lord of Formosa.* I will argue then that greed has come to the foreground as at least a major driving force to the Zhengs for all their endeavors while they operated as a Company-State in the Ming/Qing empire, just like the VOC had done to the Republic. In this way, the Zheng clan and the VOC seem worthy adversaries with greatly similar operational models, who were "into direct competition" against each other.[27]

This process then of the VOC being moved by its drive for profits through trade of porcelain and silks from China and sugar, deer skins, and venison from Taiwan, while continuously having to deal with the zest for expansion of empire by the Zheng clan, can be argued to be driven by decisions motivated by corporate greed. And this was an unsustainable process where the expansion of agriculture in Taiwan resulted in great loss of biodiversity and food storage for the people of Taiwan, as I will discuss in more detail later. Therefore, Taiwan has served as the playing field where this interaction of early corporate greed from both parties has shaped its social, cultural, and ecological landscapes to take form for years to come after.

READING THE TEXT: *LORD OF FORMOSA*

Like every historical novel, *Lord of Formosa* balances on the border where fiction and reality meet. Bergvelt first wrote her dissertation for Chinese Studies at Durham University on the historical subject of the occupation of Taiwan by the VOC. When she received the comments from her supervisor that "it read 'like an exciting novel,'" the idea was born to rewrite her dissertation into a drama while ensuring the historical facts would be correct and that the characters would be based on historical figures.[28] What makes the book a novel is her attempt "to breathe life into the characters by giving them a voice in order to tell the story from their perspective," as she explains in the Author's Note.[29] The novel focuses on two main characters: Coxinga, the heir to the Zheng clan, and the last VOC Governor of Taiwan, Frederik Coyett, even though Coyett enters the book in its second half only. The story describes the build-up toward the final conflict between Zheng the father and Coxinga in 1662, and the representatives of the VOC on the other hand, starting with Coxinga's birth in 1624 (as the first chapter) which happened to be the same year that the VOC came to Taiwan.

Ironically, it was not the initial intention or goal of the VOC to act as an agricultural colonizer to Taiwan and start a landscaping revolution, even though it can be argued that Batavia officials may have had early plans to start agriculture in Taiwan as of 1625.[30] The VOC originally only wanted to monopolize trade with China through negotiating trade rights with the Imperial Court and having a presence on land there, just like the Portuguese had in Macau.[31] The VOC only turned to the exploration of cash-cropping of first sugarcane and later also rice in Taiwan when Zheng the elder forced the VOC to stop direct trade across the Strait and make him the exclusive agent for any such trade in 1628.

It is at this juncture when the narrative of *Lord of Formosa* is first drawn into this historical re-imagination. The reader meets Zheng Zhilong when he decided that the VOC had become too much trouble to his trade. He made a move against the VOC by declaring war upon them basically forcing the VOC to enter into further trade negotiations with him. And he was right. Governor Putmans of the VOC in Taiwan decided to try to negotiate a peaceful trade deal and responded by sending a flotilla of six ships to the islands of Penghu, one of the Pescadores (an archipelago not far from the west coast of Taiwan), where Zheng Zhilong had his main offshore base. The missive held a request not to start a war but continue to work together as before, with the added bonus for Zheng Zhilong that the VOC would supply him with weapons for his fight against the Manchus, subject to a tiny condition that he would help negotiate trading rights for the VOC with the Chinese Emperor.

> Zheng's brow furrowed as he read the contents, inhaling sharply between his teeth. "They want to negotiate," he said, the tone of his voice ominously low. The old general chuckled. "Of course," he said. "That is to be expected, they are a stubborn people. What do they want?" "They expect me to act as intermediary with the Imperial Court to arrange for trading rights. They wish to have their own trade enclave, just like the Portuguese. The mere thought of it." "They are blinded by their stubbornness and greed. These people do not understand the concept of face. And I have the mandate of heaven to fight them." "War it is, then," the other man said, laying down his chopsticks across the top of his bowl. Zheng nodded, standing lost in thought for a brief moment before striding out of the building, into the darkness.[32]

Without further notice, Zheng's fleet attacked the VOC ships and sank two of them before they could break the blockade and run back to Taiwan. This forced the VOC to negotiate with Zheng on his terms and they did as they returned to the Pescadores, where Zheng Zhilong addressed the Governor as follows:

> "I am a businessman, Mr. Putmans. As you know, I have a fleet at my disposal and I have valuable contacts in all of the important ports of China. I know what products you seek and I can obtain them for you at a good price. For a small fee I would be willing to trade on behalf of your company. As your exclusive agent, of course," he added with a hint of a smile. He watched as Putmans nearly choked in consternation even before the interpreter had finished speaking, his expression betraying his feelings. . . . "*Verdomde piraat*! Damned pirate!" he heard him mutter to Van Deurzen. "Who the hell does he think he is? For the life of me we cannot allow some greedy Chinaman interfering in our business," he whispered, so the interpreter wouldn't hear. But Zheng had heard, and he observed the foreigners with cool amusement, enjoying himself.[33]

The Governor had forgotten that Zheng Zhilong had also worked as their interpreter and had an intricate knowledge about the Dutch language and culture. Zheng knew right there and then that he had been victorious. The VOC signed the peace agreement with him. And with its trading activities with China now exclusively dealt with by Zheng, suddenly the VOC had to rethink its goals for its presence on Taiwan and was forced to turn its look inland. Zheng Zhilong and the VOC both accused each other of being greedy and did not like it. Neither of them could have foreseen that their mutual disliking of each other would create a historical model of Greedflation in the seventeenth century.

When the VOC established itself in Taiwan, Taiwan was an island that had been called home by Indigenous peoples for more than 6,000 years before. It is generally considered that they started the Austronesian diaspora from Taiwan into the Pacific and beyond and that they had extensive trading networks with other peoples in the region. The landscape of Taiwan is dominated by sprawling mountain ranges with more than 260 peaks over 3,000 meters high that run from north to south, leaving a fairly narrow stretch of flat land on its West Coast that is available for development of urbanization and agriculture, with the strip being never any wider than 50 kilometers.[34] The first written record of the number of people living in Taiwan would be the report from a VOC observer in 1623, who estimated the number of Chinese living on the West Coast at "around 1500," which estimate, according to Tonio Andrade, is concurred with by most scholars.[35] This was followed by a census that was done by the VOC in 1650 which estimated the number of assimilated Indigenous peoples to be between 64,000 and 68,000.[36] In contrast, Taiwan currently is one of the countries with the highest population density in the world. The World Population Review (WPR) indicates that almost 24 million people live on 36,197 square kilometers of land. This high density ranks Taiwan as number two in the world after Bangladesh, of all countries where more than 10 million people live (World Bank 2021 data). The vast majority of people live in three major cities that are located on the west side land strip: Taipei/New Taipei City in the north, Taichung in the center, and Kaohsiung in the South. These cities are much more densely populated than the country-wide average: WPR shows that the population density in Taipei is more like 10,000 per square kilometer. It is on this West Coastal Plane that also historically most people in Taiwan have lived, starting from the arrival of the VOC at a small tongue of land over what was once called the Tayouan Bay, which is now part of Tainan City in the south of Taiwan.

Before the VOC settled on Taiwan, "there was no intensive agriculture, and therefore settlement patterns simply reflected the hunting, fishing, gathering, and small-scale farming activities of diverse native tribes."[37] This changed dramatically by VOC intervention with the intensive hunt of deer and the introduction of cash-cropping for sugarcane and rice that the VOC started, at the eventual loss of "striking biodiversity" and an "almost total elimination of all natural vegetation and wild animals from lowland areas."[38] When Zheng Zhilong forced the VOC to give up on direct trade with China, the VOC decided to introduce large-scale agriculture into Taiwan in the first instance sugarcane, and later also rice. The VOC knew that the local Indigenous peoples would not want to work the fields as their tradition was to grow crops merely to feed their own families, not for sale.[39] So they recruited farmers from China to come over to Taiwan and grow the crops on the land that the VOC would lease to them. The VOC asked Chinese businessmen to place signs in coastal cities in China: "Come to Taiwan, and the Dutch East India Company will provide land, four years of freedom from taxes, and guaranteed payments for rice and sugar."[40] As the province of Fujian was hit by drought, to many poor and starving Fujianese farmers, this must have sounded like heaven. In the decades to follow, many thousands of them crossed the Straight so that by 1645, there were at least 15,000 of them living and working in Southwestern Taiwan.[41] The sugarcane cultivation on Taiwan started when the VOC noticed that the price of sugar from China was rising considerably because of their own demand[42] so they decided to begin to cultivate sugarcane in Taiwan in 1632.[43] A letter of 1634 from Governor Putmans to the VOC leadership in Batavia mentioned a trial harvest from the previous year in a place they called Saccam (or *Chi Kan*), which would be 1633.[44] In further such letters, it was written that the VOC in fact provided small loans and company oxen to plow the land. It was such a big success, that again Governor Putmans wrote to the *Heeren XVII*: "The sugar here will be just as white as that of China, and perhaps better."[45] In 1633, a new species of sugarcane was imported by the VOC into Taiwan from China, called "Bamboo Sugarcane." This variety was more suitable for mass cultivation in Taiwan because it was abler to withstand the strong winds caused by seasonal typhoons due to its smaller and stronger stem.[46]

Sugar had become one of the hottest commodities all over Europe since its intense process of vulgarization from the mid-sixteenth to the seventeenth century.[47] Where Antwerp previously had been a center for trade of sugar, in the early seventeenth century it had moved to Amsterdam, which shows from its monopoly of sugar refining being ceded "to the profit of the United Provinces, especially Amsterdam,"[48] with refineries increasing in number there from "three refineries in 1605 to twenty-five in 1622, forty in 1650, and fifty or sixty in 1661."[49] The popularity of such an exquisite and expensive product and the substantial increase in production of refined sugar in the Republic was possible as "the ability to purchase and consume sugar, like Chinese porcelain or other luxury products, was no longer limited to the bourgeoisie because seamen, artisans, and even workers in the United Provinces earned some of the best salaries in Europe."[50] Additionally, the refined sugar was exported to other places, such as Tuscany, where "Dutch sugarloaves were preferred for their whiter, more brilliant consistency, harder, sweeter, and silkier than Venetian sugar."[51] It can hardly be surprising considering this background that the VOC Governor of Taiwan turned to the cultivation of sugarcane to develop a profitable activity on Taiwan after the Zheng trade block and his policies were aimed at facilitating sugar production in Taiwan.[52]

Sugarcane as a wild plant variety already existed in Taiwan before the VOC turned towards cash-cropping it, "because the environmental condition was very suitable for growing the plant."[53] Such conducive natural circumstances surely contributed to the success of the sugarcane production, but it is also essential that the VOC threw all its efforts in meeting the production goals. For example, in 1639 then Governor Burch banned the eating of sugarcane grown in Saccam and also ordinated that farmers were only allowed to sell the harvest to the VOC and not to the Chinese middlemen.[54] These efforts resulted in sugar production of an impressive 1.5 million kilograms per year.[55] Such were the combined inputs from the VOC, the initial tenants (*pachter* in Dutch,) and later also landowner Chinese farmers, that Andrade calls this a process of "Cooperative Colonization" or "co-colonization."[56] The VOC provided the land and favorable financial conditions to the Chinese farmers and they in turn provided the labor and skills, and also brought with them materials for farming and sugar refining. Yet, even though, the VOC gave out free lands to Chinese leaders/middlemen, the farmers were not able to make an easy profit, especially in the beginning years. The VOC would fixate how much sugarcane and other crops was mandatory to be planted and the VOC would require the farmers to sell the crops exclusively to the VOC for a fixed price, leading some farmers to complain about the expenses incurred to prepare the land and set up a sugar mill.[57] Still, the VOC seized the opportunity offered by the Chinese civil war between the Ming emperor and the Manchus to recruit even more Chinese farmers to Taiwan by waiving taxes for Chinese farmers, which policy resulted in about 14,000 farmers fleeing to Taiwan trying to avoid the war in 1649 alone.[58]

As we have learned, around the time in the civil war Zheng Zhilong made a crucial decision to switch sides from being a Ming loyalist to joining the Manchus in 1646 and thereby greatly influenced the role that Coxinga would play in this historical process. In *Lord of Formosa*, we read how he tries to explain himself to his son:

> "Don't you dare tell me of my duty, son," Zhilong said, and then lowered his voice. "Think! Look around you! Emperor Longwu is no more than a pawn in a game of chess! The Ming has been weakened for decades. You have studied the annals of history. Dynasties have come to an end before. Haven't you seen the signs?" He hesitated before he went on. "I have been in contact with their generals," he whispered. "Their generals . . . "? Zheng Sen asked. "Yes. They are willing to make me viceroy of Fujian and Guangdong if I cooperate." Zheng Sen stared at his father with open mouth. "No!," he shook his head. "You cannot mean this. I do not believe it." "We've tried, Zheng Sen. Heaven knows we have tried," his father continued matter-of-factly. "But this time, there is no stopping the enemy." His voice softened somewhat as he saw his son's expression. "There is no stopping destiny. It is only a matter of time. We have to think of ourselves, our future, or we will lose everything. We have to be practical."[59]

Coxhinga cannot believe his father's words until he suddenly sees him for the man who he really is:

> Trembling with shock, he saw his father as the man who had once removed him from his mother's side against his will. It was as if a veil had been removed from before his eyes. He now saw Zhilong's growing greed, and remembered the dark rumors that he had heard about him, the inconsistencies and lingering doubts that could never be dispelled. Somewhere from deep within him, he recalled the taunting remarks of fat Yoshi, who had called his father a pirate.[60]

And when Coxinga breaks the news to his mother, she also tries to grasp the truth about her husband's mentality:

> She shook her head in disbelief. She had long known that her husband was a selfish man, and the years of their separation had made him even more so. [. . .] But now she had the uncomfortable feeling that something else had changed. She had long suspected that he was not of such noble character as she thought when she married him, and often wondered at his increasing lack of ethics when doing business. How had he really earned his fortune? During the many years that had passed, his greed had won over his scruples—she knew that now. But this?[61]

Zheng Zilong ended up being held in detention by the Manchus until his death in 1661. From this moment on, the Zheng clan trade imperium was run by Coxinga, who managed to expand their fortune despite the civil war. And even though he may have disapproved of his father's opportunistic change in favor of the Qing rulers, Bergvelt chooses to highlight the scenario in which Coxinga abhors his father's piracy. It did not stop him from running business operations in a similar way as his father had done. The fact that the VOC had initiated the large scale agricultural development in Taiwan seemed again an opportunistic way, which had benefited Coxinga in the end. Through its developmental policies, the VOC basically did the ground work that allowed Coxinga later to maintain his huge armies and naval fleets from Taiwan.

The policies of the VOC to recruit Chinese farmers for sugarcane cultivation into Taiwan caused a major shift in the demographic landscape of Taiwan for centuries to come. By the end of the VOC rule in 1662, more than 100,000 Chinese farmers had settled in the Southwest Coastal Plain of Taiwan.[62] They were no longer fishermen or adventurers, like the few Chinese who had used Taiwan as a base for livelihood before that. These were settlers who had decided to make a permanent livelihood out of farming and who also started families by wedding local Indigenous women. They were a massive influx of Fujian people from a culture foreign to either the Indigenous peoples or the Dutch and they brought in their ways of farming, housing, and village construction that would go on to become the so-called "Taiwanese" tradition. And this development was only strengthened by the Zheng family rule after they took over Taiwan, and later by the new Qing Dynasty beyond that.

The change of the use of the land naturally impacted the landscape of the region. The Chinese middlemen landlords were eager to promote rice farming next to sugarcane, so extensive rice fields and complementary irrigation structures were developed. From 1645 to 1661, the land used for farming had increased from 3,00 *morgen* to 12,000 *morgen*.[63] In order to expand the agricultural use of the land, trees were cut and massive deforestation followed, which in turn destroyed the natural habitat of deer in that region. Together with the intensive hunt of deer for their skins and meat, this led to a serious decrease in deer population.[64] These changes in landscape deeply affected the traditional livelihoods of the Indigenous peoples, who had depended on fishing, hunting and gathering in the forested landscape for centuries and who found their ways of life seriously threatened by this development, while the VOC simultaneously regulated and restricted (Indigenous) hunting and trading of deer and their products through a system of licensing for the benefit of earning levies.[65] Additionally, the VOC promoted the import and breeding of oxen and cows for use as farming animals to increase productivity in the sugarcane and rice fields, which resulted in the domestication of these animals in Taiwan. The number of cows increased from about 20 in the late 1620s to 300 by 1635.[66]

Even though the original intent of the VOC was to catch the heights of the sugar trade and promote the cultivation of sugarcane in Taiwan, the VOC also tried the growing of other crops like indigo, mulberry, cotton, ginger, and rice.[67] Based on the (sub) tropical weather system, crops like sugarcane and rice flourished best so that the VOC and the Chinese farmers naturally focused on those two crops, where sugarcane was grown mostly for export and rice mostly for domestic consumption.[68] Yet, as sugarcane was a trade priority of the VOC, taxation and land use policies were favoring the cultivation of sugarcane and this resulted at times in the situation that domestic prices for rice were very high or supply was limited. Additionally, not all sugarcane could be processed into sugar for sale, because at times, rice supply was not enough. For example, in 1639, it was reported by the Governor of Taiwan in a letter to Batavia, that "farmers sold unprocessed sugarcane to the hungry, since 'the poor man can hardly find any other food,'" while at the same time, the VOC ordinated that raw sugarcane could no longer be sold directly for consumption so that it would be all available for refinement.[69] Also, the VOC would raise the price they would pay for sugar against that for rice, in order to prevent the Chinese from switching completely to rice.[70] It was reported by the VOC in 1651 that due to the VOC land and taxation policies, many Chinese had fallen into poverty.[71]

The narrative of greed as laid out in *Lord of Formosa*, is not limited to the VOC. In fact, the aforementioned Chinese middlemen also did their best to enrich themselves, in their turn at the cost of the Indigenous peoples. The story reflects on this as three aboriginal village elders come to see the Governor of the VOC:

> "Please be informed that our people are finding it increasingly difficult to make a living. The wealthy Chinese leaseholders, those who have the exclusive rights to trade in our villages, are setting the prices for their goods at too high a price." The village elder glanced uncomfortably at Ting-bin, who was a fair representation of the type of wealthy Chinese merchants he spoke of. Ting-bin was far more than an interpreter: he was also a company tax collector, village head, leaseholder, and a successful businessman who had his connections, both on Formosa and in China. He spoke Chinese, Dutch, and a smattering of Portuguese, which made him invaluable to the company. In fact, he personified all that the aboriginal islanders had come to despise. "You see, Excellency, the number of deer has been falling at an alarming rate. As you know, trading deer products is our only means to survive, but more villages are forced to share the same hunting fields. And if this goes on . . . " The man paused trying to find the right words. "The Chinese merchants, they set unfairly low prices for buying

our products; we have no choice but to accept. They are making it impossible for many of our people to get by."[72]

It was in the end the missionaries who had come to Taiwan to christen and educate the Indigenous peoples, who had warned the VOC of the disaffection brewing among them:

> It was Antonius [one of the local missionaries, ed.] who brought the growing discontent among the native islanders to Frederic's [Coyett, Assistant Governor, ed.] attention. Before Daniel's departure, he and Antonius had tried to persuade [Governor] Verburgh to take the necessary measures to curb the Chinese merchants' power and protect the villagers from their greed. The two missionaries stood far closer to the aboriginal community than any other company employee, and they knew firsthand what kind of impact the trade leases had on their lives. They could see up close how unfair the monopolies created by the leasehold system were for the aboriginals, who were left with no other alternative than to sell their goods to the Chinese merchants at the low prices they were offered.[73]
>
> At Frederic's urging, Verburgh finally called a meeting with the island's seven major leaseholders, among them village headman Guo Wen-bao. Verburgh suggested to the Chinese they set the same prices for their goods across the island, explaining that the inflation that they were imposing on the islanders was becoming a source of discontent.[74]

As the Chinese merchants resisted this proposal, Verburgh then declared the introduction of weekly free-markets, where goods could be sold and bought by anybody freely. These markets were so successful, that the Chinese leaseholders complained to the VOC and announced that next term they would no longer bid at the auctions for exclusive village trading rights. The VOC realized that this refusal would seriously undermine their profit-making and as a result the free markets were cancelled and the difficult situation for the villagers was aggravated:

> In the meantime, the company filled its coffers with the money raised from the leasehold auctions, collected hoofdbrief taxes, hunting licenses, fishing licenses, and sales taxes on everyday essentials. Long gone were the days when migrant farmers had been given tax breaks and even subsidies to lure them to the island. Now the livelihoods of the farmers were at the mercy of their Dutch creditors and tax collectors. Added to this were the corrupt soldier-school teachers and other low-ranking company officials who posed as inspectors and squeezed money from the people to add to their wages.[75]

The VOC drew income not only from trade to and from Taiwan, but also from levying taxes: "the most important was a tax of 10% of rice harvests, but there were others, such as taxes on livestock slaughtering, on weighing and measuring, and on the export of venison. The right to collect these taxes was sold in annual auctions, usually to Chinese."[76] Another important tax that the VOC introduced was the head-tax (*hooftgelt*), a residency fee which each Chinese resident in Taiwan had to pay.[77] The most substantial local source of income for the VOC, however, was the auctioning of various licenses to Chinese middlemen, including licenses for deer hunting and, most importantly, the exclusive licenses to trade with Indigenous villages.[78] By structurally using middlemen systems for the gathering of income, the VOC clearly tried to maximize profits while keeping cost low through maintaining a basic presence in Taiwan only. While the income of the Chinese farmers stayed low and in a year of a bad harvest could not support their families, the VOC made profit margins in the Republic on sugar from Taiwan between 262 percent and 423 percent.[79] The difficult circumstances for the farmers and especially the discontent caused by the system of the head tax and its collection by the VOC soldiers, even brought them to revolt against the VOC in 1652. This event was called the Guo Huan-yi Rebellion and it was suppressed by use of great force by the VOC.[80]

In *Lord of Formosa*, this uprising is being described elaborately. It may well have been the turning point in the fate of the VOC on Taiwan, the beginning of the third Act where events start to take the narrative towards the grand finale of the story: the siege by Coxinga and his armies of the VOC's Taiwan stronghold Fort Zeelandia and the end of VOC rule in 1662. The Chinese farmer rebellion is said to have raised in arms at least one third of all Chinese against the VOC and is even rumored to have been instigated, or at least supported, by Coxinga himself who increasingly seems to have been orchestrating the demise of the VOC and its policies of Greedflation behind the scenes. Guo Huan-yi as local rebel leader spurs on his men:

> "We toil in their fields to grow their rice and their sugar. For this, we are paid next to nothing! And when we buy the rice and sugar at the markets, we pay taxes for what we ourselves have planted and harvested. With our very own sweat and blood!" He paused for his words to take effect. "Lord Koxinga, the great lord of the imperial surname, fights with his troops in China against our enemy, the Manchus. He has instructed us to fight our foreign enemies here, on Formosa. It is our duty!" He paused again as the crowd roared in approval. "The Hollander soldiers and officials harass and abuse us. They take from us the last of our money in their greed and corruption. They have cleverly pushed many of us into debt so that they can control us like slaves. We can no longer accept this! We've had enough! Enough!"[81]

The rapid development of agriculture in Taiwan under the VOC and the obvious importance of its potential for food supply sourcing had also been noticed by Coxinga who had decided that Taiwan would be his next stop. In July 1656, Coxinga sent an edict to the VOC that he claimed sovereignty over all the Chinese people on Taiwan while banning all traffic across the Taiwan Strait and ordering Chinese residents to return to China: the next stage of the war of greed had been initiated.[82] On February 1, 1662, after a siege of almost 10 months, Frederik Coyett was forced to surrender Taiwan to Coxinga. Coyett was allowed to leave Taiwan with all the VOC ships and surviving soldiers, and to take enough provisions to reach Batavia. As he arrived in Batavia sometime later, the word of the appearance of his flotilla with its flagship the *'s Gravelande* proudly ahead, reached Batavia's Governor-General Maetsuijker, who rushed to the window and immediately realized what had happened as he saw the majestic warship with lowered sails, ready to enter the harbor. The *'s Gravelande* had been stationed in Taiwan for many years and its arrival here could only mean one thing: the rumors had proven true and Taiwan had been lost:

> He lowered himself back onto his chair. He removed his glasses shakily and stared ahead, feeling older than before. He rubbed at his temples, where the tension had been building up over the past few days but where he now had a splitting headache. He was painfully aware that he was partly to blame for all this. But was any of this really his fault? Could he have prevented this? For months the *Heren Seventien* [or *Heeren XVII*, ed.] had put him under a great deal of pressure. Those wealthy, mighty men, respected by all and residing in their stately homes on the canals of cities such as Amsterdam, Rotterdam, and Delft. None of them had any idea how things worked in the Far East; a few had never even been to Asia. They didn't understand why the Chinese refused so persistently to grant the VOC the trade that it so badly wanted. But they would not listen. The only thing that they wanted was results.[83]

THE TRANSCULTURAL AND SUSTAINABILITLY PERSPECTIVE

Four hundred years later, Taiwan looks very different from how it had been when the VOC first settled here. Urbanization has become the dominant trend in both habitat and livelihood formation while rural villages suffer diminished economic activity and a strongly aging demography. Agriculture has come to represent only a small percentage of all gross income and socially speaking, farming has fallen into marginalization. This is a significant shift from the more "traditional" economic organization that emerged from the VOC era, which had been agricultural, and that lasted all the way into the late 1980s when a hyper-fast process of industrialization resulted in what has become known as Taiwan as one of the Asian Tigers.

It has become obvious from the analysis so far that many of the elements that have defined the agricultural landscape in Taiwan as traditional, have been first introduced into Taiwan by or as a consequence of the policies of the VOC that were implemented for the purpose of raising income out of trade from Taiwan, especially of sugarcane. As I have mentioned, these elements include the introduction of major crops like sugarcane and rice, the import of oxen to facilitate the plowing, the immigration of farmers from China to work the fields, and very importantly, also the process itself of large-scale agriculture through wet paddy field farming and irrigation structure development. These elements then have shaped the look of agriculture in Taiwan for many centuries and have as such inerasably marked the landscaping of Taiwan. Wherever one travels in Taiwan in the right season, the lush bright green of new shooting in wet paddy fields shines through, even sometimes in urban areas, where sparse agricultural fields are fighting for survival. Demographically, the organized immigration of Han Chinese in the seventeenth century is considered the first and constitutive settlement of the Chinese in Taiwan, paving the road for many more to follow. Through intermarrying with the local Indigenous women of the West Coastal Plane, the Han Chinese influx also caused the local Indigenous population in that area to more or less fully integrate into this demographic movement. Politically, the growth of the Chinese community on Taiwan and the wish to settle in Taiwan that this may have aroused in Coxinga, caused the new Qing Dynasty to take an interest in Taiwan as a possible territory of China for the first time in history. And very importantly, ecologically, the landscape has suffered from these developments through large-scale deforestation and loss of bio-diversity.

Even though at first glance, it may seem surprising that a lot of people now living in Taiwan would consider a cultural product like sugarcane and its cultivation as something very Taiwanese while it is clearly a foreign element in origin, this is in fact a very common phenomenon that is explained very well by the theory of transculturality as first proposed by the German philosopher Wolfgang Welsch.[84] Welsch argues that contrarily to the concept of mono-culture as defined by nation, borders, or sense of tribe, culture does not form and has not formed through history in a closed-off manner, as separate spheres and/or bound within borders and languages. Cultural formation is a process that transcends such borders and limitations and as such is a transcultural process. In this perspective, cultural formation is not merely the result of continuous intermixing of cultural elements, native and foreign, and thus a result of hybridity as proclaimed by Homi Bhabha[85] and Edward Said.[86] According to Welsch, the intermixing of cultural elements becomes a prerequisite for "transculturality" and will create such a state of transcultural awareness when people lose the sense of distinction between ownness and foreignness of a certain cultural element. For example, some Taiwanese may adopt the habit of drinking espresso coffee, but they will be very much aware that it is a foreign cultural element. On the other hand, when talking about sugarcane cultivation in Taiwan, the locals will consider this as a native cultural element while in reality it was foreign-imported by the VOC. This means that espresso coffee drinking is a hybrid cultural element and sugarcane is a transcultural one, as the Taiwanese have lost their sense of distinction between ownness and foreignness. It is such a process of reconstruction of the historical mixing of cultural elements that will establish the form of transculturality.[87] And I have argued in this chapter that, through the historical analysis of the environmental history of sugarcane landscaping in Taiwan and by reading selected parts of *Lord of Formosa*, this process bears the mark of transculturality in Taiwan.

Another important perspective that flows from this historical treatise and re-imagination of the creation of the sugarcane landscape in Taiwan is that it was in fact an unsustainable development under the current Sustainable Development Goals (SDGs). As I have argued, the development by the VOC of large-scale agriculture for sugarcane and rice in the Southwestern Coastal Plane was not sustainable because of its negative consequences for the landscaping, the residents, and the endemic wildlife. Biodiversity was seriously affected through deforestation, re-shaping of the land to facilitate cash-cropping and over-hunting of mostly deer. This would in modern terms constitute an issue under SDG 15, the call to protect life on land and restore terrestrial eco-systems. The motive of profiteering that drove the cash-cropping development resulted in the proclamation by the VOC of policies of pricing, lending, and taxing that favored sugarcane cultivation for the purpose of trade. This process led to situations where domestic food supplies were insufficient or at least unstable and to inflationary prices for basic foods like rice. As was shown, there were even times when local farmers had to resort to the consumption of raw sugarcane for lack of staple food. Such conditions would certainly qualify as irresponsible and unethical business development under SDGs 8, 9, and 12 and would also trigger concerns under SDG 2 on zero-hunger. The issue that is mostly relevant for the argument of this chapter however, is the immense inequality of wealth that was created by the VOC and the Zheng clan while pursuing the corporate greed, through the practices of landscape re-shaping for agriculture and trade which is currently a major issue under SDG 10.

The reading of *Lord of Formosa* as featured in this chapter facilitates this discussion because the author has marked a distinct pathway through the main historical and narrativized events of the trade war between the VOC and the Zheng clan by using the words *greed*, *greedy*, and even *inflation*, which are the main indicators of motive that move corporations in their active participation in the process of world-making that throughout history, has an impact on the lives of all sentient beings on this planet. In this way I have set out to follow the thread that has been laid down in the narrative of the novel to guide the reader through the maze of the re-imagination of this environmental historic event.

Thus, I bring forth this transcultural and sustainable reading of the environmental history of this landscaping development by the VOC in Taiwan to evidence that in fact, *Greedflation* may be a new word invented to fit our economic times of the early 2020s, but not a new phenomenon as Edwards had mentioned. The VOC had as the first corporation in history, already developed such practice with similar consequences in Taiwan in the seventeenth century. I argue that the reading of the novel and the re-imagination of the historic environmental events narrated therein on current terms can help us to form a more concrete picture of how unsustainable practices like corporate greed and *Greedflation* will impact our current and future life world. In that sense this exploration is indeed a transcultural analysis of the historical events. I hope that through this contribution to the debate on sustainable consequences of landscaping over time, we can avoid unsustainable development traps and better envisage sustainable solutions to our current planetary anthropogenic emergency.

NOTES

1. *Fortune*, https://fortune.com/2023/04/05/end-of-capitalism-inflation-greedflation-societe-generale-corporate-profits/, last viewed April 6, 2023.

2. Isabelle M Weber and Evan Wasner, "Sellers' Inflation, Profits and Conflict: Why Can Large Firms Hike Prices in an Emergency?" ScolarWorks@UMassAmherst, 5. Last viewed July 1, 2023.

3. Bernard Looney, as quoted in the *Daily Mail*, 2 November 2021, https://www.thisismoney.co.uk/money/markets/article-10157957/BP-like-cash-machine-claims-boss-price-oil-gas-soars.html, last viewed July 5, 2023.

4. Grist.org, 18 February 2023, https://grist.org/economics/bp-exxon-shell-backing-off-climate-promises/, last viewed July 5, 2023.

5. The Guardian, 12 March 2023, https://www.theguardian.com/commentisfree/2023/mar/12/the-guardian-view-on-corporate-greed-its-causing-inflation, last viewed July 5, 2023.

6. BEA, https://www.bea.gov/data/income-saving/corporate-profits, last viewed July 5, 2023.

7. The second novel, titled *Commandeur van de Kaap*, is about the seventeenth century Dutch East India commander Simon van der Stel of Cape of Good Hope.

8. Femme S. Gaastra, "The Organization of the VOC," in *The Archives of the Dutch East India Company (VOC) and the Local Institutions in Batavia (Jakarta)*, eds.

Louise Balk, Frans van Dijk, Diederick Kortlang, Femme Gaastra, Hendrik Niemeijer and Pieter Koenders (Leiden, The Netherlands: Brill Publishers, 2007), 13–60.

9. Lodewijk Opmaat, *The World's First Stock Exchange* (New York: Columbia University Press, 2014).

10. Oscar Gelderblom, Abe De Jong, and Joost Jonker, "The Formative Years of the Modern Corporation: The Dutch East India Company VOC, 1602–1623," *The Journal of Economic History* 73, no. 4 (2013): 1050–1076.

11. Stephen R. Bown, *Merchant Kings: When Companies Ruled the World, 1600–1900* (Madeira Park, BC, Canada: Douglas & McIntyre, 2021).

12. Grist.org, 18 February 2023, https://grist.org/economics/bp-exxon-shell-backing-off-climate-promises/, last viewed July 5, 2023.

13. As quoted in Arthur Weststeijn, "The VOC as a Company-State: Debating Seventeenth Century Dutch Colonial Expansion," *Itinerario* XXXVIII, no.1 (2014): 22.

14. The word *Geoctroieerd* in the name of the VOC stems from the old French word *O(c)troi*, which in turn comes from the late Latin *auctorizare*. According to Websters's Revised Unabridged Dictionary 1913, *octroi* means: a privilege granted by the sovereign authority, as the exclusive right of trade granted to a guild or society; a concession. Websters's Revised Unabridged Dictionary, https://www.websters1913.com/words/Octroi, last viewed July 11, 2023.

15. See provisions XXXVI and XXXVII of the VOC Charter, based on the text as published by François Valentijn in his work Oud en Nieuw oost-Indiën of 1724, found and last viewed on July 11, 2023 on https://www.vocsite.nl/geschiedenis/octrooi1602/, with scans of the original text of the charter available for viewing on the Dutch National Archives website here: https://www.nationaalarchief.nl/onderzoeken/archief/1.04.02/invnr/1/file/NLHaNA_1.04.02_1_01?eadID=1.04.02&unitID=1&query=

16. Ann Heylen, "The Transnational in Taiwan History: A Preliminary Exploration," *Concentric: Literary and Cultural Studies* 36, no. 1 (2010): 14.

17. Arthur Weststeijn, "The VOC as a Company-State: Debating Seventeenth-Century Dutch Colonial Expansion," *Itinerario*, XXXVIII.1 (2014): 16.

18. Weststeijn, "The VOC as a Company-State," 15.

19. Weichung Cheng, War, Trade and Piracy in the China Seas, 1622–1683 (Leiden, The Netherlands: Brill, 2014), 37–39, 41.

20. Kenneth Pomeranz and Steven Topik, *The World That Trade Created: Society, Culture, and the World Economy, 1400 to the Present*, 2nd Edition (Armonk, NY: M.E. Sharpe, 2006), 127.

21. Robert J. Antony, "Turbulent Waters: Sea Raiding in Early Modern South East Asia," *The Mariner's Mirror* 99, no. 1 (2013): 25.

22. Tonio Andrade, Lost Colony: The Untold Story of China's First Great Victory over the West (Princeton, NJ: Princeton University Press, 2011).

23. Robert J. Antony, Pirates in the Age of Sail (New York: W.W. Norton & Company Inc., 2007), 111–114.

24. Richard Blakemore, "Review," *International Review of Social History* 59, no. 1 (2014): 149–150, reviewing: Wei-Chung Cheng, *War, Trade and Piracy in the China Seas, 1622* New York *1683* (Leiden, The Netherlands: Brill Publishers, 2013).

25. Pomeranz and Topik, *The World That Trade Created*, 127.

26. Tonio Andrade, "The Rise and Fall of Dutch Taiwan, 1624–1662: Cooperative Colonization and the Statist Model of European Expansion," *Journal of World History* 17, no. 4 (2006): 432.

27. Andrade, "The Rise and Fall," 450.

28. https://bookish.asia/author-interview-joyce-bergvelt/, last viewed July 19, 2023.

29. Joyce Bergvelt, *Lord of Formosa* (Manchester, UK: Camphor Press, 2018), 446.

30. Tonio Andrade, *How Taiwan Became Chinese: Dutch, Spanish, and Han Colonization in the Seventeenth Century* (New York: Columbia University Press, 2008), chapter 6, 6 of 18, paragraph 17.

31. Hui-wen Koo, "Sugar Production and Trade in Dutch Colonial Taiwan," *Taiwan Historical Research* 28, no. 2 (2021): 46.

32. Bergvelt, *Lord of Formosa*, 76.

33. Bergvelt, *Lord of Formosa*, 80.

34. Ronald Knapp, "The Shaping of Taiwan's Landscapes," in *Taiwan: A New History*, ed. Murray A. Rubinstein (Armonk, NY: M.E. Sharpe, 1999), 6.

35. Andrade, *How Taiwan Became Chinese*, chapter 6, 13 of 18, note 5.

36. Jennifer A. Liu, "Aboriginal Fractions: Enumerating Identity in Taiwan," *Medical Anthropology: Cross-Cultural Studies in Health and Illness* 31, no. 4 (2012): 334.

37. Knapp, "The Shaping of Taiwan's Landscapes," 11.

38. Knapp, "The Shaping of Taiwan's Landscapes," 8.

39. Andrade, "The Rise and Fall," 430.

40. Andrade, "The Rise and Fall," 430.

41. Andrade, "The Rise and Fall," 430.

42. Koo, "Sugar Production and Trade," 46.

43. Hua-Pi Tseng, "Sugarcane and the Environment under Dutch Rule in Seventeenth Century Taiwan," in *Environmental History in the Making: Volume II: Acting*, eds. Cristina Hoanaz de melo, Estelita Vaz, and Ligia M. Costa Pinto (New York: Springer Publishers, 2017), 192.

44. Andrade, "The Rise and Fall," 437.

45. Andrade, "The Rise and Fall," 437.

46. Tseng, "Sugarcane and the Environment," 192 and 194.

47. Eddy Stols, "The Expansion of the Sugar Market in Western Europe," in *Tropical Babylons: Sugar and the Making of the Atlantic World, 1450* New York *1680*, ed. Stuart B. Schwartz (The University of North Carolina Press, 2004), 240.

48. Stols, "The Expansion," 272.

49. Stols, "The Expansion," 273–274.

50. Stols, "The Expansion," 273–274.

51. Stols, "The Expansion," 273–274.

52. Tseng, "Sugarcane and the Environment," 192.

53. Tseng, "Sugarcane and the Environment," 191.

54. Tseng, "Sugarcane and the Environment," 193.

55. Tseng, "Sugarcane and the Environment," 193.

56. Andrade, "The Rise and Fall," 450. And Andrade, *How Taiwan Became Chinese*, chapter 6, 10 of 18, paragraph 27.

57. Koo, "Sugar Production and Trade," 47.
58. Tseng, "Sugarcane and the Environment," 195.
59. Bergvelt, *Lord of Formosa*, 127.
60. Bergvelt, *Lord of Formosa*, 128.
61. Bergvelt, *Lord of Formosa*, 130.
62. Tseng, "Sugarcane and the Environment," 195.
63. Koo, "Sugar Production and Trade," 72.
64. Tseng, "Sugarcane and the Environment," 196.
65. Andrade, "The Rise and Fall," 442.
66. Tseng, "Sugarcane and the Environment," 197.
67. Tseng, "Sugarcane and the Environment," 194.
68. Tseng, "Sugarcane and the Environment," 197.
69. Andrade, "The Rise and Fall," 440.
70. Koo, "Sugar Production and Trade," 54.
71. Koo, "Sugar Production and Trade," 55.
72. Bergvelt, *Lord of Formosa*, 129.
73. Bergvelt, *Lord of Formosa*, 180.
74. Bergvelt, *Lord of Formosa*, 181.
75. Bergvelt, *Lord of Formosa*, 185.
76. Andrade, "The Rise and Fall," 441.
77. Andrade, "The Rise and Fall," 445.
78. Andrade, "The Rise and Fall," 441.
79. Koo, "Sugar Production and Trade," 69.
80. Andrade, "The Rise and Fall," 445.
81. Bergvelt, *Lord of Formosa*, 199.
82. Koo, "Sugar Production and Trade," 51.
83. Bergvelt, *Lord of Formosa*, 426.
84. Wolfgang Welsch, "Transculturalitaet: Lebensformen Nach Der Aufloesung Der Kulturen," *Information Philosophie* 2 (1992): 5–20. And Wolfgang Welsch, "Transculturality: The Changing Form of Cultures Today," *Filozofski vestnik* XXII, no. 2 (2001): 59–86.
85. Homi K. Bhabha, *The Location of Culture* (London Routledge, 1994).
86. Edward Said, 1996. As quoted by Wolfgang Welsch in Wolfgang Welsch, "Transculturality in Literature: A Phenomenon as Old as It Is Current," *World Literature Studies* 3, no. 14 (2022): 5.
87. Wolfgang Welsch, "Transculturality in Literature: A Phenomenon as Old as It Is Current," *World Literature Studies* 3, no. 14 (2022): 5.

BIBLIOGRAPHY

Andrade, Tonio. "The Rise and Fall of Dutch Taiwan, 1624–1662: Cooperative Colonization and the Statist Model of European Expansion." *Journal of World History* 17, no. 4 (2006): 429–450.

Andrade, Tonio. *How Taiwan Became Chinese: Dutch, Spanish, and Han Colonization in the Seventeenth Century*. New York: Columbia University Press, 2008.
Andrade, Tonio. Lost Colony: The Untold Story of China's First Great Victory over the West. Princeton, NJ: Princeton University Press, 2011.
Antony, Robert J. Pirates in the Age of Sail. New York: W.W. Norton & Company Inc., 2007.
Antony, Robert J. "Turbulent Waters: Sea Raiding in Early Modern South East Asia." The Mariner's Mirror 99, no. 1 (2013): 23–38.
Bergvelt, Joyce. Lord of Formosa. Manchester, UK: Camphor Press, 2018.
Bhabha, Homi K. The Location of Culture. London: Routledge, 1994.
Blakemore, Richard. "Review." International Review of Social History 59, no. 1 (2014): 149–150.
Bown, Stephen R. Merchant Kings: When Companies Ruled the World, 1600–1900. Madeira Park, BC, Canada: Douglas & McIntyre, 2021.
Cheng, Wei-Chung. War, Trade and Piracy in the China Seas, 1622–1683. Leiden, The Netherlands: Brill Publishers, 2013.
Gaastra, Femme S. "The organization of the VOC." In The Archives of the Dutch East India Company (VOC) and the Local Institutions in Batavia (Jakarta), edited by Louise Balk, Frans van Dijk, Diederick Kortlang, Femme Gaastra, Hendrik Niemeijer and Pieter Koenders, 13–59. Leiden, The Netherlands: Brill Publishers, 2007.
Gelderblom, Oscar, Abe De Jong, and Joost Jonker. "The Formative Years of the Modern Corporation: The Dutch East India Company VOC, 1602–1623." The Journal of Economic History 73, no. 4 (2013): 1050–1076.
Heylen, Ann. "The Transnational in Taiwan History: A Preliminary Exploration." Concentric: Literary and Cultural Studies 36, no. 1 (2010): 9–33.
Knapp, Ronald. "The Shaping of Taiwan's Landscapes." In Taiwan: A New History, edited by Murray A. Rubinstein, 3–26. Armonk, NY: M.E. Sharpe, 1999.
Koo, Hui-wen. "Sugar Production and Trade in Dutch Colonial Taiwan." Taiwan Historical Research 28, no. 2 (2021): 45–87.
Liu, Jennifer A. "Aboriginal Fractions: Enumerating Identity in Taiwan." Medical Anthropology: Cross-Cultural Studies in Health and Illness 31, no. 4 (2012): 329–346.
Opmaat, Lodewijk. The World's First Stock Exchange. New York: Columbia University Press, 2014.
Pomeranz, Kenneth and Steven Topik. The World That Trade Created: Society, Culture, and the World Economy, 1400 to the Present, 2nd Edition. Armonk, NY: M.E. Sharpe, 2006.
Said, Edward. 1996. As quoted by Wolfgang Welsch in: Welsch, Wolfgang. "Transculturality in Literature: A Phenomenon as Old as It Is Current." World Literature Studies 3, no. 14 (2022): 5–11.
Stols, Eddy. "The Expansion of the Sugar Market in Western Europe." In Tropical Babylons: Sugar and the Making of the Atlantic World, 1450–1680, edited by Stuart B. Schwartz, 237–288. Chapel Hill, NC: University of North Carolina Press, 2004.

Tseng, Hua-Pi. "Sugarcane and the Environment under Dutch Rule in Seventeenth Century Taiwan." In Environmental History in the Making: Volume II: Acting, edited by Cristina Hoanaz de melo, Estelita Vaz, and Ligia M. Costa Pinto, 189–200. New York: Springer Publishers, 2017.

Weber, Isabelle M. and Evan Wasner. "Sellers' Inflation, Profits and Conflict: Why Can Large Firms Hike Prices in an Emergency?" UMassAmherst *Economics*, no. 343 (2023): 1–51. https://scholarworks.umass.edu/econ_workingpaper/343/

Welsch, Wolfgang. "Transculturalitaet: Lebensformen Nach Der Aufloesung Der Kulturen." Information Philosophie, no. 2 (1992): 5–20.

Welsch, Wolfgang. "Transculturality: The Changing Form of Cultures Today." Filozofski vestnik XXII, no. 2 (2001): 59–86.

Welsch, Wolfgang. "Transculturality in Literature: A Phenomenon as Old as It Is Current." World Literature Studies 3, no. 14 (2022): 5–11.

Weststeijn, Arthur. "The VOC as a Company-State: Debating Seventeenth-Century Dutch Colonial Expansion." Itinerario XXXVIII, no. 1 (April 2014): 13–44.

Chapter 2

Empire's Nature

Economics, Ecology, and Formosan Tea in Nineteenth-Century Western Travelers' Natural Histories

Li-Ru Lu

Ecocriticism is environmentally oriented literary and cultural studies; it is the study of the representation of the environment and ecological issues in the works of literature and popular culture.[1] Recent developments in East Asian ecocriticism frames postcolonial ecocritical approaches to the Asia-Pacific: Jeffrey Santa Ana's *Empire and Environment: Ecological Ruin in the Transpacific* (2022), for instance, inquiries into the imperialist-based capitalist economies wrought by the material and historical forces of the Western empire and underscores the interrelation of imperialism, the appropriation of natural resources, and the ruination of the environment in the Transpacific.[2] In *Empire and Environment*, John Charles Ryan insightfully asserts that "the history of imperialism is a history of entwined ecological [. . .] degradation."[3] To be sure, the essays in *Empire and Environment* have raised awareness about empire and environmental destruction in the Asia-Pacific. Nevertheless, in *Empire and Environment* and other recent East Asian ecocritical studies—such as Jason Paolo Telles's and John Ryan's *Environment, Media, and Popular Culture in Southeast Asia* (2022),[4] Taiwan does not receive attention from ecocritics. This chapter in turn explores some uncharted spaces of East Asian ecocriticism: it focuses on the natural histories about tea and tea plants in nineteenth-century Formosa (now Taiwan). In aligning itself to the argumentative direction of *Empire and Environment*, this chapter explores the links between nineteenth-century imperial motivations and representations of Formosan tea and tea plants; it investigates the entanglement of empire, economics, ecology in the traveling natural histories of nineteenth-century Western travelers who visited Formosa.

Formosa or Ilha Formosa, meaning the "evergreen and beautiful isle," was named by passing Portuguese navigators in the sixteenth century.[5] The Island of Formosa was colonized by the Dutch in the seventeenth century; the Spanish built a settlement in northern Formosa for a very brief period but were driven out by the Dutch in 1642.[6] In 1662, Zheng Cheng-gong, a loyal soldier of China's Ming dynasty, defeated the Dutch and established a base of operations in Formosa, but his forces were defeated by China's Qing dynasty in 1683.[7] From 1683 to 1895, the Qing empire ruled Formosa and in 1895 Taiwan was ceded to Japan.[8] Qing rule gradually collapsed during the second half of the nineteenth century; in the Opium Wars of 1839–1842 and 1856–1860, the British army easily defeated the Qing army and navy.[9] Following Qing China's defeat, Britain imposed unequal treaties on the defeated Chinese and forced the Qing to open up many ports for foreign trade, including ports in Formosa.[10] In 1860, the Qing government's ratification of the Treaty of Tientsin allowed the opening of Formosa's trading ports.[11] Before 1860, Formosa was largely unknown and unexplored by foreigners. After the opening of Formosa's Ta-kau Port, Tamsui Port, and Keelung Port for trade in 1860, plenty of Westerners visited Formosa. These travelers—such as Robert Swinhoe (1836–1877), John Dodd (1838–1907), William Hancock (1847–1914), Joseph Beal Steere (1842–1940), and James Wheeler Davidson (1872–1933)—made observations and documented the unknown landscapes, natural resources, especially tea, of Formosa. Generally written in the form of travel narratives, the writings of these Western travelers were pioneering texts that delineated Formosa's tea and other natural productions in the nineteenth century; however, these works receive little attention from ecocritics.[12] In the twenty-first century, though a few ecocritical studies—such as Simon C. Estok's *East Asian Ecocriticisms*, Karen Laura Thornber's *Ecoambiguity*, and Chia Ju Chang's and Scott Slovic's *Ecocriticism in Taiwan*—began to explore literary works about Korea, Japan, and Taiwan,[13] these books still overlook nineteenth–century Westerners' writings about Formosa. This chapter, by contrast, focuses on the traveling natural histories of five late-nineteenth-century Western travelers—Swinhoe, Dodd, Hancock, Steere, and Davidson; it explores the historical and cultural underpinnings of issues surrounding tea in nineteenth-century Formosa and examines how these Westerners represented Formosa's tea (such as tea trade, tea plantation, or tea plants). Specific questions addressed are: How might the texts on nineteenth-century Formosan tea relate to traveling natural histories? What are the intricate links between late-nineteenth-century Western travelers' imperial motivations and their representations of Taiwanese tea? How did Swinhoe, Dodd, Hancock, Steere, and Davidson report their observations of Formosa's tea or tea trees in the late nineteenth century? Do any of the writings about nineteenth-century Formosan tea articulate environmentalist concerns?

ECONOMICS, ENVIRONMENTAL REPRESENTATION, AND TRAVELING NATURAL HISTORY

Before the mid-nineteenth century, Formosa was little known and mostly unexplored by foreign travelers. After Formosa's ports had been opened for foreign trade by the Western powers in 1860,[14] a great number of English and American travelers visited Formosa. Through their writings, these visitors—Swinhoe (a British naturalist who worked as the first consul in Formosa), Dodd (an English merchant who promoted Taiwanese tea to America), Hancock (a British botanist who worked in Tamsui Maritime Customs from 1881 to 1882), Steere (an American naturalist who traveled in Formosa and the Far East), and Davidson (a naturalist and the first American consul in Taiwan)—provide a plethora of natural history information on Taiwan's natural resources and environmental productions, particularly tea. In 1864, Swinhoe's essay, "Notes on the Island of Formosa," was published in the *Journal of the Royal Geographical Society of London*.[15] Like Swinhoe, Dodd also traveled the Island of Formosa. He described Taiwan's natural productions and his conduct of tea trade in "Formosa," an article that appeared in *The Scottish Geographical Magazine* in 1895.[16] Touring Tamsui and its neighboring regions in 1881,[17] another British traveler, Hancock detailed Formosa's tea plants and tea trade in an essay titled "Tamsui Trade Report. For the Year 1881"; this essay appeared in *China Imperial Maritime Customs. Reports on Trade at the Treaty Ports, For the Year 1881*.[18] Steere, another Western traveler, made a visit to Formosa in 1871 and 1873;[19] he authored *Letters from Formosa* and *Formosa and Its Inhabitants*.[20] Steere also published a scholarly article titled "Formosa," which appeared in *Journal of the American Geographical Society of New York* in 1874.[21] Davidson—another American visitor to late-nineteenth-century Formosa—traveled throughout Taiwan, recording his observations of Formosa's natural resources in *The Island of Formosa: Past and Present. History, People, and Commercial Prospects. Tea, Camphor, Sugar, Gold, Coal, Sulphur, Economical Plants, and Other Productions*; this book urged American investment in Taiwan and Davidson's efforts helped America to replace Britain as the major Western trading nation in Formosa.[22] In *The Island of Formosa,* Davidson closely delineated Taiwan's tea trade and tea districts in "Chapter XXIII: The Formosan Tea Industry."[23]

Composed as travel narratives, the aforementioned works of Swinhoe, Dodd, Hancock, Steere, and Davidson are works of natural history. Natural history was generally used to refer to the study of all aspects of the environment; it recorded information related to all environmental productions, including tea. Natural history draws on scientific knowledge about nature, and in this sense, its "scientific bent" is a characteristic feature of natural history.[24] However, natural history is also a kind of belles-lettres, for the genre documents a world captured by human observers in a particular cultural frame.[25] As both science and belles lettres, natural history is hence a "narrative practice" of humans (particularly Europeans) observing natural phenomena and representing the environment.[26] Since many natural history writers—such as Hancock and Steere—combine their inquiry into nature with a narrative of their journeys, it is expected that traveling natural history provides an eyewitness account of the environment encountered during one's travels.[27]

In the late nineteenth century, the Western travelers who visited Formosa—such as Dodd, Davidson, Archibald Ross Colquhoun (1848–1914), James H. Stewart-Lockhart (1858–1937), and Herbert J. Allen[28]—frequently described the natural resources they encountered during their travels and provided important early records of Formosa's environment. In their works, these writers consistently report the abundant natural resources of Formosa. Accordingly, their accounts reveal continuing connections between economic interests, environmental representations, and traveling natural histories. Alerting the outside world to the commercial possibilities of the island of Formosa, numerous nineteenth-century accounts of Taiwan served as promotional writings in which the primary concern was the economic or utilitarian dimension of nature. The ensuing passage from Colquhoun's and Stewart-Lockhart's "A Sketch of Formosa" offers an example:

> Like all countries shrouded in mystery, Formosa was credited with fabled treasures of the earth, the field, and the forest. It was to be an EL Dorado for the merchant.[29]

Showing the obviously commercial purpose of Colquhoun's and Stewart-Lockhart's visit to Formosa, this passage reveal continuing connections between economic interests, environmental representations, and imperialist ideas of Formosa. Through their appropriation of nature as an exploitable resource or capitalist commodity, imperial motivations and economic interests entered Colquhoun's and Stewart-Lockhart's environmental discourse.

The report on Formosa's abundance of natural resources and environmental productions—such as tea—can also be found in the following excerpt from Herbert J. Allen's "Notes of a Journey through Formosa from Tamsui to Taiwanfu":

> The information obtained from time to time about Formosa showed that everything there was in its infancy. Rice, camphor, wheat, coffee, tobacco, tea, and sugar were all grown there; and no doubt other tropical produce would thrive, if there was a good government and colonists were encouraged to settle. As far as Europeans were concerned, however, he might say of it, as the Irishman said of Ireland, that it was the finest if not the healthiest country in the world—to live out of.[30]

Alerting the outside world to the economic potential of Formosa, numerous nineteenth-century accounts of Formosan tea and other natural productions served as promotional writings in which the primary concern was the utilitarian or commercial dimension of nature. Seeking to exploit economic possibilities in Formosa, the British consular officer Allen felt the country to be rich in potential export items. Attempting to entice readers with the prospects of Taiwan's tea and other export items, Allen's traveling natural history placed emphasis on the capitalist and imperial aspects of the representation of nature.

The traveling natural histories by nineteenth-century Western travelers who visited Formosa are deeply embedded in "capitalist-intensive western economic forces," as Richard H. Grove observes in *Green Imperialism: Colonial Expansion, Tropical Island Edens and the Origins of Environmentalism, 1600–1860*; these texts shared a common economic narrative regarding the natural environment they represented.[31] As Timothy Sweet explains in his essay "Economy, Ecology and Utopia in Early Colonial Promotional Literature," promotional writings articulated fundamental perceptions of the environment and natural resources, while encoding complex ecological and economic theories.[32] Promising new possibilities for harnessing the plenitude of nature, Formosa invited foreign travelers to delineate their natural assets—especially tea—and to develop modes of economy based on new networks of resource and information exchange. In their writings, Dodd, Hancock, Davidson, and most nineteenth-century traveling natural historians expressed the colonial desire to know and to exploit the land. Colonial powers, as Elizabeth DeLoughrey, Jill Didur, and Anthony Carrigan postulate in *Global Ecologies and the Environmental Humanities: Postcolonial Approaches*, "render the landscape into resources to be owned cultivated, or simply extracted."[33] The traveling natural histories of Swinhoe, Dodd, Hancock, Steere, and Davidson, in a word, attest to the interdependent relationship between colonial greed and representations of Formosan tea.

The connections among economics, environmental representations, and traveling natural histories can also be found in Hancock's "Tamsui Trade Report. For the Year 1881":

> The best Formosa Oolongs are considered better than any Oolongs grown on the mainland, and the prices obtained are a proof of their high appreciation by consumers. Out of the 12,000,000 lbs. to 13,000,000 lbs. exported, only from 500,000 lbs. to 1,000,000 lbs. go to England, where these Oolongs are mixed with other Teas, to flavour them. The taste for Formosas alone has not yet been acquired in England, whereas in the States the aromatic and pungent flavour seems to suit the American palate. The future Tea trade of Formosa looks very promising, for the export has increased in a most wonderful way.[34]

Hancock wrote this report when he worked at Tamsui Maritime Customs in the 1880s; this report is related to Britain's overseas capitalist expansion. Emphasizing foreign mercantile interest, this excerpt alerts Britain to the commercial dimensions of Formosa and reveals the historical embeddedness of "Formosa Oolongs" in the British imperial enterprise. Asserting that "The future Tea trade of Formosa looks very promising, for the export has increased in a most wonderful way," Hancock in this passage made explicit the otherwise implicit connections between the traveling natural histories on Formosan tea, economic interest, and colonialism. Revealing the desire to exploit Formosa Oolong in nineteenth-century Taiwan, these lines echo Edward Said's notion of Orientalism. Orientalism, as has been formulated by Said, connotes the "attitude of nineteenth-century and early-twentieth-century European colonialism"; it is a "corporate institution" empowered to deal with the environment "by making statements about it, describing it," knowing it, and finally "ruling over it."[35]

REPRESENTING FORMOSAN TEA: THE COLONIAL DIMENSION OF ENVIRONMENTAL REPRESENTATIONS IN NINETEENTH-CENTURY WESTERNERS' TRAVELING NATURAL HISTORIES

In the nineteenth century, Formosa was largely uncharted territory. The opening of Formosa to foreign trade in 1860, however, attracted traveling naturalists and merchants to report on Formosa. In this context, Swinhoe, Dodd, Hancock, Steere, and Davidson alerted the outside world to the physical nature and the commercial possibilities of Formosa. Venturing into Formosa for the exploration of tea and other natural products, these Western naturalists delineated Formosa's geological landscape and articulated the economic prospects of Formosa's natural resources, especially tea. In *Formosa and Its Inhabitants*, for instance, Steere closely delineated the tea trade and tea plants of Formosa:

> The tea of Formosa is becoming of great importance commercially, especially to Americans, since the kind produced—Oolongs—is principally consumed in the United States. It is cultivated toward the north end of the island only and on high, rolling lands that cannot be irrigated and that appear to be generally cultivated by hand, the soil being dug up with rude hoes and spades to prepare it for planting. The soil is a rich red clay, and the tea tree and camphor grow native upon it. The tea plants are set out five or six feet apart, and are allowed to grow three or four feet in height, and are usually quite thick and bushy from the frequent picking. The young leaves just opening are the only ones used for tea. These, after being dried by the Chinese cultivators, are taken to Tamsui, when the foreign merchants have all the apparatus for preparing the tea for shipment.[36]

Combining the scientific language of exact measurement with personal observation and containing important geological information about the plantation of Oolong tea in Tamsui,[37] this quoted text provides a rare early account about Formosan tea and offers unique, first-hand information to an audience anticipating further elaboration, discovery, and trade. Steere in this passage introduced readers to the physical character of northern Formosa's tea plantation and accessibly and eloquently communicates scientific knowledge about nature. This quoted extract exhibits the prominent economic concerns within Steere's mode of environmental representation. Economics entered Steere's ecological understanding as nature came to be classified under the category of "commodity." These lines reveal the intimate linkage between economics and the environment in Steere's traveling natural history. Interestingly enough, in this passage Steere also recognized that "The soil is a rich red clay, and the tea tree and camphor grow native upon it": his observations invoke an awareness of interconnectedness and hint at a notion of wholeness; this quoted material exhibits Steere's perception of natural systems and complex interdependencies existing among soil, tea tree, and camphor tree in a given region. Steere's natural historical account here, in brief, displays his sensitivity to the interconnectedness of nature and to interrelationships in a natural community and thus exhibiting his proto-ecological vision.[38]

In effect, Steere was not the only traveling naturalist who journeyed to Tamsui and described the Formosan tea in nineteenth-century northern Taiwan; his contemporary Swinhoe and Dodd also journeyed to the area. In *The Island of Formosa*, Davidson states,

> Robert Swinhoe may be called the discoverer of the Formosan Tea industry, and John Dodd the promoter.[39]

It is Swinhoe who discovered the commercial and capitalist potential of Formosan tea trade in Tamsui and reported his discovery to the British government.[40] An accomplished naturalist, Swinhoe was a member of the Royal

Geographical Society of London.[41] Most of his writings were published in *Journal of the Royal Geographical Society of London*. His writings for the Royal Geographical Society included new discoveries about Formosan tea. When he was in Taiwan, Swinhoe took notes about potential commercial products, such as tea. As a result of his understanding of Formosa and his enthusiasm for the island, Swinhoe was appointed in December 1860 as the first British consular official in Formosa and a commercial representative of British interests.[42] In "Notes on the Island of Formosa," Swinhoe included references to certain commodities for export trade from Tamsui. He cataloged the natural resources found in Tamsui: "The exports from Tamsui comprise the following articles: Rice, indigo, coarse sugar, jute, ground-nut cakes, camphor, coal, grass-cloth fibre, wood, rattans, tea, rice-paper pith, pickled vegetables, small pulse, barley, wheat, and sulphur."[43] This passage catalogs the items of export trade in Tamsui and reveals the economic dimensions of the Formosan environment. After mentioning Tamsui's trade products, Swinhoe in "Notes on the Island of Formosa" commented separately on the major products of the region. His comments on tea are worthy of particular attention:

> The tea grown near Tamsuy was not of superior quality, but I have been informed, on the decision of three tea-tasters to whom I sent samples, that it would readily find a market in Australia, the Cape, and Singapore. It rates at a price of 10 dollars a picul, and is much imported by Chinese dealers at Amoy and Foochow, to mix with the better class of teas; and the mixed commodity is then sold to foreign merchants [. . .] As the hills, however, are no great distance from the harbour, this could be improved by energetic speculators.[44]

In 1863, Swinhoe read the above-quoted information at a meeting of the Royal Geographical Society in London.[45] His report on Formosan tea soon caught the attention of John Dodd, a Scottish traveling naturalist and merchant. Born at the time of heightened European exploration activities, Dodd travelled to Taiwan for his own overseas adventures. He took great interest in the Taiwanese tea trade and commenced building up that product in the northern areas of Formosa. Though Swinhoe in "Notes on the Island of Formosa" remarked that the commercial potential for Formosan teas was ordinary,[46] Dodd found a tremendously successful career in the tea trade. In the end, he established an unexpected market for his teas in America. In a section titled "Commerce and Industry" in the article "Formosa," Dodd narrated how growing tea plants in North Formosa brought unprecedented prosperity for him:

> While exploring the camphor districts in 1865, I came across cinnamon and wild tea plants, the latter being ten or twelve feet high. I got down a fair amount of cinnamon to the coast, but found that the expense of sorting the outer and inner barks and packing them for shipment to England was too great. With tea I was more successful. On making inquiries I found [. . .] small patches of tea were cultivated in the gardens of farmers, but that it was grown principally for home consumption. All the tea I could get I bought up, and finding that it fetched good price in Macao, I at once made loans to the farmers, through my *compradore*, for the purpose of extending the cultivation, and also imported slips of the tea plant from Amoy. I then started firing on a small scale at Bangka, and afterwards took larger premises at Twa-tu-tia. In the course of three for four years Formosan teas acquired a reputation in America, and before the end of sixties in one year I had dispatched in one year two sailing vessels to America, one with a full cargo and the other nearly full. Others followed my example, and the exports increased by bounds year after year [. . .] I have no statistics showing what is the present quantity of tea exported from Tamsuy, but I should say that it is certainly less than 500,000 chests, equal to about 20,000,000 pounds. The hills and soil of North Formosa are excellently adapted for tea-growing, and, indeed, for almost any crop.[47]

It was Dodd who made Taiwanese tea known world-wide. Dodd's successful tea enterprise won him the title—the "Father of Taiwanese Tea." As indicated by this passage, Dodd promoted the economic prospects of North Formosa's natural productions and emphasized that Taiwan was rich in potential export items, especially tea and crops. Displaying humanity's resistance to recognizing the intrinsic value of nature, Dodd's writing stresses the importance of the commercial aspects of nature in representations of this era. Once again, an economic motivation swayed Dodd's understanding of nature as a desirable commodity or valuable product. The quotation exhibits the intrinsic overlays between economics and the environment in Dodd's promotional natural history.

The report on Formosa's economic interests and commercial possibilities can also be found in the concluding paragraph from Dodd's "Formosa":

> Tea is now by far the most important article of export [. . .] In conclusion, I must say that British merchants were the first foreigners to [. . .] open up the coal trade [. . .] and were the first to start the tea trade, which has grown to such an enormous extent, and that therefore the credit of having laid the foundations of the prosperity of Formosa is principally due to them.[48]

Seeking to exploit commercial possibilities in Formosa, the British tea merchant John Dodd felt the country to be rich in two potential export items—tea and coal. Dodd in these lines lays great stress on the significance of the Formosan tea trade, maintaining that tea industry is the foundation of "the prosperity of Formosa."[49] Attempting to entice readers with the prospects of Formosa's tea trade, Dodd's promotional/traveling natural history placed emphasis on the economic aspects of natural representation. Through the valuation of nature as a resource or product, economics entered Dodd's environmental discourse. This quoted passage, in sum, exhibits the fundamental linkage between economics and the environment in the writings of early naturalists in Formosa. Dodd's writing attests to the interdependence of colonial prerogatives and narrative accounts of Formosan lands, revealing nineteenth-century Western travelers' desires for natural resources and showing the largely economic characterization of the Formosan environment in most Westerner travelers' minds.

REPRESENTING FORMOSAN TEA: THE PROTO-ECOLOGICAL DIMENSION OF ENVIRONMENTAL REPRESENTATIONS IN NINETEENTH-CENTURY WESTERNERS' TRAVELING NATURAL HISTORIES

While nineteenth-century English and American travelers in Formosa were fully cognizant of the commercial and mercantile value of tea, a few among them—particularly Steere, Hancock, and Davidson—also provided early environmentally sensitive accounts about Formosan tea; these accounts frequently delineate the abundance and beauty of Formosa's natural productions. The ensuing passage from Steere's *Letter from Formosa* serves as a distinct example:

> I made several trips about Tamsui [. . .] visiting some large tea plantations, the tea being planted upon the hills, where rice cannot be grown for lack of water. The tea plant, from continual picking, is kept down to a low, thick bush from two to four feet high, and it is planted in rows about three feet apart each way. The leaves are a very dark green, and those plantations that are in a flourishing state are beautiful, the plants having the large white, fragrant flowers scattered over them.[50]

The natural historical language of Steere here contains not only important scientific information about Tamsui's "large tea plantations" but also Steere's closely-observed descriptions of Formosan tea plants: the color, size, and smell of Tamsui's tea leaves and tea flowers—all contribute to the beauty of Taiwanese tea. Combining his travel narrative with the scientific description of nature, these lines vividly convey the charm of tea plants and tea flowers in Tamsui, express Steere's ingrained sense of awe and his admiration for the green world, and thus display his proto-ecological sensibility. The term "proto-ecological sensibility" was coined by Michael P. Branch in "Indexing American Possibilities" to mean "incipient ecological sensibilities," such as an appreciation of nature, the expression of ecological awareness," the celebration of the beauty of nonhuman nature, and the "sensitivity to [the] interconnectedness" of nature and to interrelationship in a natural community.[51] Steere's natural historical accounts about Formosan tea, in a word, display his proto-ecological sensibility and his enthusiasm for the beauty of Formosan tea plants.

In effect, the expression of proto-ecological sensibility and the environmentally sensitive accounts about Formosan tea can also be found in Hancock's "Tamsui Trade Report. For the Year 1881":

> The Tea plants themselves are bushy and luxuriant, averaging from 2 to 3 feet high, but some are certainly from 4 to 6. The bark of the trees and the stones in the vicinity are covered copiously with moss and lichens, bearing testimony to the moistness of the climate. In the Formosa Tea plantations, little attention appears to be paid to drainage.[52]

In this quoted extract, Hancock delineates the abundance and charm of the "bushy and luxuriant" tea plants in Tamsui. More importantly, he represented the broader environment in which these tea plants were located—"the trees and the stones" in the surrounding area, "the moistness of the climate," the "moss and lichens" on the "bark of the tress."[53] It is a suggestive version of an ecosystem, "a model of interrelatedness in nature," as Donald Worster calls it.[54] Hancock's natural historical account about the Formosan tea-district here hints at the notion of the web of nature, especially the interconnections between Formosa's tea plants and the nonhuman nature around the tea-plantation area—such as moss, lichens, and the trees and stones "in the vicinity," thus revealing his awareness of environmental relationships and proto-ecological sensibility.

Interestingly enough, the proto-ecological sensibility and the awareness of environmental relationships also appear in Steere's natural historical text titled "Formosa":

> The culture of tea has but just commenced, but has already reached considerable importance in Tamsui. The kind mostly cultivated is Oolong, which goes most of it to the United States. The cultivation can be carried to any extent in the north of the island, as there are great plains of high rolling lands lying from Changwha to Tamsui, of the rich red clay which is so favorable for tea. These high plains are the home of the camphor tree, many trunks and stumps of which still lie over the ground, and of the wild tea-tree, of which I saw one specimen, with the trunk a foot in diameter and thirty feet high. The inhabitants gather the wild tea, but it is of bad quality, either by nature or from being badly cured. These high

> plains are now lying desert, as they cannot be irrigated, while tea is one of the few plants that needs no irrigation.[55]

These lines present the environmental knowledge and scientific information about the "home" of the tea-trees in nineteenth-century Taiwan. Like the previous extract from Hancock's "Tamsui Trade Report," this excerpt exhibits Steere's proto-ecological awareness of relationships and interconnectedness within natural systems. Such an awareness displays a significant multispecies ethnographic quality in Steere's natural historical accounts: the notions of interrelatedness and the entangled quality of life on earth.[56] In *The Book of Nature*, Margaret Curzon Welch contends that the accounts of the lives of nonhumans should contain "every scrap of available evidence on habitat and geographical distribution of a species";[57] her emphasis is on looking at plants, birds, and animals within their habitats and not as isolated creatures.[58] Such an emphasis suggests a multispecies ethnographic vision of the world, for Welch's ideal life accounts allow the focus to be placed upon a natural species and all its interrelations with the rest of the natural world. In this quoted passage, Steere narrated his discovery and his observation of "the home"—the habitat—of Taiwanese tea-trees, informing his readers that "the great plains of high rolling lands" are the home of both tea-tree and camphor tree and that "the rich red clay" is "so favorable for tea,"[59] thereby displaying his proto-ecological and multispecies-ethnographic sensitivity to the interconnectedness of nature and to interrelationships in a natural community. Remarking on the connection existing between "the rich red clay" and tea-tree and especially the connection existing between Formosan "high plains," "the camphor tree," and the tea-tree, Steere presciently suggests the notions of interrelatedness and of species interdependence; this excerpt thus anticipates the focus on this natural interrelatedness of future multispecies ethnographers. Focusing on multiple species, multispecies ethnographers emphasize the webs of interspecies dependence.[60] Multispecies ethnography is a kind of life writing grounded in attentiveness to the lives of nonhumans, and multispecies ethnographic studies, as S. Eben Kirksey and Stefan Helmreich postulate, "impinged on notions of relatedness" and of species interdependence.[61] Steere's accounts about Taiwanese tea-tree here, in this vein, become pioneering multispecies ethnographies.

In addition to the sensitivity to interconnectedness of nature and to interrelatedness in a natural world, a few English and American travelers in Formosa—particularly Hancock and Davidson—also perceived the deforestation and the ruination of the natural environment in early Formosa. The following passage selected from Hancock's "Tamsui Trade Report. For the Year 1881" serves as a distinct example:

> Everywhere one is struck by the amount of new ground which is being broken up in order to set more Tea. The first thing is to clear away the long grass and brushwood, cut down what trees there may be (I speak more particularly of the hills on the borders of the savage territory), stub up the roots, make bonfire [. . .] When this plant has run for a while, it is replaced by Tea.[62]

Stating that the people in Tamsui "cut down what trees there may be" "in order to set more Tea,"[63] Hancock in this passage reported and described humankind's thoughtless destruction of nineteenth century Tamsui's natural resources. Like Hancock, Davidson was also aware of the destruction of forests and woods in nineteenth-century Taiwan; in *The Island of Formosa*, Davidson's keen environmental awareness of the devastation of Formosa's timber and forest trees is noticeable:

> Formerly the savages in the extreme north were but rarely disturbed, and the boundaries of their districts remained the same year after year; but now, with the vision of ample profits from the Tea trade, the colonists made great progress in annexing the "hill barbarians'" land, and within ten years (1868 to 1878) the greater part of two ranges of hills, varying in elevation from 1,000 to 4,000 feet, lying to the east of Taipehfu, had been cleared of all heavy timber and forest trees, and the jungle with its savage occupants had been replaced by little plantations of Tea [. . .] Tea merchants were also attracted by the increased trade, and were willing to make cash advances to prospective growers, thus assisting in the laying out of plantations. [. . .] In preparing virgin soil in the hill districts, the long grass and shrubbery are first removed, the trees are then felled and the roots dug out [. . .] The jungle being thus removed, a crop of sweet potatoes is planted, to be replaced, the following season, with young tea plants.[64]

The Island of Formosa exhibits Davidson's insightful awareness of the immense environmental changes that were already being worked on the land in late-nineteenth-century Formosa; in this book Davidson showed that the Formosan landscape itself had been profoundly transformed. Davidson in these quoted lines delineated the colonists' massive mistreatment of nature. This quoted passage reveals Davidson's ecological worries and environmental anxiety, thus displaying an incipient awareness of the immense destruction of tree species occurring late-nineteenth-century Formosa. To be sure, Hancock's and Davidson's sense of environmental awareness is not, strictly speaking, ecocentric. Yet their sensitivity to Formosa's ecological ruin and environmental destruction is an essential precursor to the idea of environmental conservation. Davidson's and Hancock's awareness of the devastation of forests and trees in Formosa, in this vein, is the combined product of their proto-ecological sensibility. This quoted extract from *The Island of Formosa*, in brief, reveals Davidson's deep concern regarding Formosan forests' critical endangerment. Addressing the ecological implications of the disappearance of timbers, jungles and forests in Taiwan, these lines exhibit Davidson's insightful awareness of the immense environmental changes and ecological degradation that were already being worked on the hills and high mountains in late-nineteenth-century Formosa.[65] Hancock's and Davidson's accounts of human-induced environmental degradation is significant culturally and historically for tracking ecological changes in early Formosa. More than one hundred and fifty years ago, Davidson was concerned about humankind's mistreatment of nature, about the relationship between humans and the environment, and about environmental damage: he was aware of the loss of richness of Formosan "heavy timber and forest trees" in inland mountains.[66] Davidson's astute observations of the drastic changes being worked on forest trees and in Formosan "high plains" and jungles are among the earliest expressions of ecological worries, anxiety, and concern over environmental transformations: his fear for the devastation of nature.

CONCLUSION

In "Afterword: Ecocritical and Literary Futures," the concluding chapter of *East Asian Ecocriticism*, Karen Laura Thornber argues that ecocriticism has to uncover "environmental subtexts from a range of creative works" and to open up the existing textual archives that had previously been neglected in literary and cultural studies.[67] Certainly, nineteenth-century natural historical texts about Formosan tea are an integral part of those overlooked "environmental subtexts" and textual archives.[68] The traveling natural histories of Swinhoe, Dodd, Hancock, Steere, and Davidson have not received the critical attention they deserve. They are still largely unheard of even today, remaining unknown and neglected in the field of East Asian ecocriticism. Focusing on the aforementioned authors, this chapter has examined some unexplored spaces of nineteenth-century ecocriticism in order to retrieve obscured voices that articulate incipient ecological sensibilities and even nascent multispecies ethnographic sensibilities, including an appreciation of nonhuman nature, a sensitivity to interconnectedness and interrelationships among species in natural environments, and the sense of environmental awareness about the devastation of nature.

As faithful and reliable reporters on Formosa and its tea industry at the threshold of the twentieth century, Swinhoe, Dodd, Hancock, Steere, and Davidson also offer compelling accounts of Formosan tea and thus alerted readers to the commercial possibilities there. They made explicit their vision of the imbrications between the natural historical texts on tea and colonial economic interests in plant resources, especially tea. Embodying the colonial and capitalistic desire to delineate, "know," and exploit the land, the traveling natural histories of Swinhoe, Dodd, Hancock, Steere, and Davidson attest to "the colonial/imperial underpinnings of environmental practice"—as Graham Huggan and Helen Tiffin have put it[69]—in nineteenth-century Formosa Island and echo Said's notion of Orientalism. As formulated by Said, Orientalism is "a system of knowledge about the Orient" and "a Western style" for describing, "dominating, restructuring, and having authority over the Orient."[70] Indeed, the remarkable accounts of Swinhoe, Dodd, Hancock, Steere, and Davidson not merely contain vividly written representations that offer knowledge about Formosa and its tea but also attest to the inextricable interrelationship between textually natural historical representations of East Asia and colonial desire in the nineteenth century.

NOTES

1. Karen Laura Thornber, "Afterward: Ecocritical and Literary Futures," in *East Asian Ecocriticism: A Critical Reader*, eds. Simon C. Estok and Won-Chung Kim (Palgrave Macmillan, 2015), 237; John Charles Ryan, "Beyond East Asian Landscapes: A Survey of Ecocriticism in Thai Literary Studies," in *Referred Proceedings of the International Conference on Trends in Economics, Humanities and Management*, eds. Jiri Strouhal and Michel Plaisent (Pattaya, Thailand, 2015), 26–30, accessed May 8, 2023, http://icehm.org/upload/3350ED815006.pdf.

2. Jeffrey Santa Ana, Heidi Amin-hong, Rina Garcia Chua, Zhou Xiaojing, eds. *Empire and Environment: Ecological Ruin in the Transpacific* (University of Michigan Press, 2022), 1. In addition to Jeffrey Santa Ana's *Empire and Environment*, Philip Aghoghovwia's essay titled "Postcolonial Nature" also explores the interrelationship between imperialism and the destructive extraction of natural resources motivated by capitalist and economic growth; see Philip Aghoghovwia, "Postcolonial Nature," in *Nature and Literary Studies,* eds. Peter Remien and Scott Slovic (Cambridge University Press, 2022), 211–228.

3. John Charles Ryan, "If We Return, We Will Learn: Empire, Poetry and Biocultural Knowledge in Papua New Guinea," in *Empire and Environment: Ecological Ruin in the Transpacific*, eds. Jeffrey Santa Ana, Heidi Amin-hong, Rina Garcia Chua, and Zhou Xiaojing (University of Michigan Press, 2022), 94–110.

4. Jason Paolo Telles, John Charles Ryan, and Jeconiah Louis Dreisbach, eds., *Environment, Media, and Popular Culture in Southeast Asia* (Springer, 2022).

5. George Williams Carrington, *Foreigners in Formosa 1841–1874* (University of Michigan Press, 1973), 3. Having arrived in Far Eastern waters in the sixteenth century, the Portuguese named Taiwan "Ilha Formosa." However, there is no evidence of the Portuguese's actual settlement in Formosa.

6. Carrington, *Foreigners in Formosa*, 3–4.

7. Carrington, *Foreigners in Formosa*, 4–5.

8. Qing China rule over Formosa ended when Taiwan was ceded to Japan by the Treaty of Shimonoseki in 1895. From 1895, the Japanese began its colonization of Formosa. Taiwan was a part of the Japan empire under Japanese colonial rule from 1895 to 1945.

9. One main contributing factor in Qing China's decline was European imperialism. In the late nineteenth century, Europe's leading countries—primarily Britain and France—exerted their control over large portions of Asia and Africa, putting pressure even on the traditional superpower of East Asia, imperial China. The most devastating blow came in the Opium Wars of 1839–1842 and 1856–1860, after which Britain imposed unequal treaties on the defeated Chinese.

10. Carrington, *Foreigners in Formosa*, 21–22.

11. Carrington, *Foreigners in Formosa*, 22.

12. Published in 1999, Harold M. Otness's *One Thousand Westerners in Taiwan, To 1945: A Biographical and Bibliographical Dictionary* began to recognize the significance of Western travelers who visited Formosa from the sixteenth century to the early twentieth century. Yet this book is presented mainly as a "biographical and bibliographical dictionary" and, therefore, is unable to examine nineteenth-century Westerners' works in depth and detail.

13. Simon Estok and Won-Chung Kim, eds., *East Asian Ecocriticisms: A Critical Reader* (Palgrave Macmillan, 2013); Karen Laura Thornber, *Ecoambiguity: East Asian Literatures and Environmental Crises* (University of Michigan Press, 2015); Chia Ju Chang and Scott Slovic, eds., *Ecocriticism in Taiwan: Identity, Environment, and the Arts* (Lexington Books, 2015).

14. Carrington, *Foreigners in Formosa*, 129.

15. Robert Swinhoe, "Notes on the Island of Formosa," *Journal of the Royal Geographical Society of London* 34 (1864): 6–18, https://rdc.reed.edu/i/2fee8170-ee11-4de6-b5d4-fc09a4cf36c6.

16. John Dodd, "Formosa," *The Scottish Geographical Magazine* XI (1895): 570, https://rdc.reed.edu/i/784ca96e-71ff-4443-a8f2-752a748a9a62.

17. Otness, *One Thousand Westerners in Taiwan*, 70.

18. William A. Hancock, "Tamsui Trade Report. For the Year 1881," in *China Imperial Maritime Customs. Reports on Trade at the Treaty Ports, For the Year 1881* (Statistical Department of the Inspectorate General, 1882), 1–38, https://rdc.reed.edu/i/07fa4929-760a-4544-84e3-4347bd952875.

19. Otness, *One Thousand Westerners in Taiwan*, 150.

20. Joseph Beal Steere, *Formosa and Its Inhabitants*, ed. Paul Jen-Kuei Li (Institute of Taiwan History, Academia Sinica, 2002).

21. Joseph Beal Steere, "Formosa," *Journal of the American Geographical Society of New York* 6 (1874): 302–334. https://rdc.reed.edu/i/4cd68be9-1612-45fb-b739-95012d7b80e7.

22. Otness, *One Thousand Westerners in Taiwan*, 39.

23. J. W. Davidson, "Chapter XXIII: The Formosan Tea Industry," in *The Island of Formosa: Past and Present. History, People, and Commercial Prospects* (Macmillan & Co., 1903), 371–376. In effect, *The Island of Formosa* is the major English

Language survey of Taiwan for its days and still the most frequently consulted English language source about early Taiwan.

24. Ashton Nichols, "Natural History," in *Nature and Literary Studies,* ed. Peter Remien and Scott Slovic (Cambridge University Press, 2022), 123; Patrick D. Murphy, *Farther Afield in the Study of Nature-Oriented Literature* (University Press of Virginia, 2000), 45.

25. John Tallmadge, "Toward a Natural History of Reading," in *The ISLE Reader: Ecocriticism, 1993–2003*, eds. Michael P. Branch and Scott Slovic (University of Georgia Press, 2003), 291–292.

26. Elizabeth M. DeLoughrey, Renee K. Gosson, and George B. Handley, *Caribbean Literature and Environment: Between Nature and Culture* (University of Virginia Press, 2005), 5.

27. Traveling natural history, in this sense, is both travel writing and nature writing.

28. The dates of Herbert J. Allen's birth and of his death are unknown. However, it is certain that Allen was an English Consular officer in Tamsui in 1873 and he wrote the consular trade reports for Northern Taiwan for the year 1873–1874; See Otness, *One Thousand Westerners in Taiwan*, 3.

29. A. R. Colquhoun and J. H. Stewart-Lockhart, "A Sketch of Formosa," *The China Review* 13 (1885): 188, https://rdc.reed.edu/i/2fb0712d-3ff6-4a71-acfc-1fded48fd6d3.

30. Herbert J. Allen, "Notes of a Journey through Formosa from Tamsui to Taiwanfu," *Proceedings of the Royal Geographical Society of London* 21 (1877): 266, https://rdc.reed.edu/i/6dbd3295-77e2-4448-bdd7-e2abda3387e3.

31. Richard H. Grove, *Green Imperialism: Colonial Expansion, Tropical Island Edens and the Origins of Environmentalism, 1600–1860* (Cambridge University Press, 1995), 2.

32. Timothy Sweet, "Economy, Ecology, and Utopia in Early Colonial Promotional Literature," *American Literature* 71, no. 3 (1999): 399–401.

33. Elizabeth DeLoughrey, Renee Gosson, and George Handley, *Caribbean Literature and Environment: Between Nature and Culture* (University of Virginia Press, 2005), 12.

34. William A. Hancock, "Tamsui Trade Report," 7.

35. Edward Said, *Orientalism* (Pantheon, 1978), 2, 3, 6.

36. Steere, *Formosa and Its Inhabitants*, 129.

37. Steere in this passage exhibited his ability to combine the language of exact measurement with personal observation. One reason for this is that Steere was a product of his age. While the science-literature split may exist today, there was no such split in nineteenth-century America. In his essay, Robert J. Scholnick argues that in eighteenth-century and nineteenth-century America, science and literature had not yet separated. See Robert J. Scholnick, "Permeable Boundaries: Literature and Science in America," in *American Literature and Science*, ed. Robert J. Scholnick (University Press of Kentucky, 1992), 7.

38. In next section, this chapter will closely examine the proto-ecological dimension of environmental representations in the traveling natural histories of Steere, Hancock, and Davidson.

39. Davidson, "Chapter XXIII: The Formosan Tea Industry," 373.

40. Davidson, "Chapter XXIII: The Formosan Tea Industry," 373.

41. Carrington, *Foreigners in Formosa*, 161.

42. Carrington, *Foreigners in Formosa*, 129.

43. Swinhoe, "Notes on the Island of Formosa," 15.

44. Swinhoe, "Notes on the Island of Formosa," 16.

45. Carrington, *Foreigners in Formosa*, 129, 135.

46. Swinhoe, "Notes on the Island of Formosa," 15.

47. Dodd, "Formosa," 569–570.

48. Dodd, "Formosa," 570.

49. Dodd, "Formosa," 570.

50. Joseph Beal Steere, Letters from Formosa, published in *Ann Arbor Courier* (January 2, 1874), 25, https://rdc.reed.edu/i/e8947d6f-a3a5-4439-a321-7e54032ea723.

51. Michael P. Branch, "Indexing American Possibilities: The Natural History Writing of Bartram, Wilson, and Audubon," in *The Ecocriticism Reader: Landmarks in Literary Ecology*, eds. Cheryll Glotfelty and Harold Fromm (University of Georgia Press, 1996), 288–291.

52. Hancock, "Tamsui Trade Report," 27.

53. Hancock, "Tamsui Trade Report," 27.

54. Donald Worster, *Nature's Economy: A History of Ecological Ideas* (Cambridge University Press, 1994), 378.

55. Steere, "Formosa," 316–317.

56. Thom Van Dooren and Deborah Bird Rose, "Lively Ethnography: Storying Animist Worlds," *Environmental Humanities* 8, no. 1 (2016): 77–94.

57. Margaret Welch, *The Book of Nature: Natural History in the United States 1825–1875* (Northeastern University Press, 1998), 5.

58. In *The Book of Nature*, Welch explores the accounts about the lives of nonhuman nature; she focuses her emphasis on looking at animals, plants, and birds within their habitats. Such an emphasis echoes the multispecies ethnographic notions of many twenty-first-century multispecies ethnographic scholars—particularly Mara de Genaro, Deborah Bird Rose, Thom Van Dooren, Eben Kirksey, and Stefan Helmreich—who place a fresh emphasis on the subjectivity of nonhuman creatures whose lives are unavoidably connected and interdependent with "other kinds of living selves" and with the rest of the natural world. See S. Eben Kirksey and Stefan Helmreich, "The Emergence of Multispecies Ethnography," *Cultural Anthropology* 25, no. 4 (2010): 545–576; Deborah Bird. Rose, "Ethnography," in *Keywords: For Environmental Studies*, eds. Joni Adamson, William A. Gleason, and David N. Pellow (New York University Press, 2016), 110–111.

59. Joseph Beal Steere, "Formosa," 316–317.

60. Kirksey and Helmreich, "The Emergence of Multispecies Ethnography," 553.

61. Kirksey and Helmreich, "The Emergence of Multispecies Ethnography," 550.

62. Hancock, "Tamsui Trade Report," 26.

63. Hancock, "Tamsui Trade Report," 26.

64. Davidson, "Chapter XXIII: The Formosan Tea Industry," 379–380.

65. The notions of ecological degradations and environmental damages appear in Karen Laura Thornber's *Ecoambiguity: East Asian Literatures and Environmental*

Crises (University of Michigan Press, 2015), 3, 91–93; these notions also appear in Stefania Barca's essay titled "History," an article in *Keywords: For Environmental Studies*, eds. Joni Adamson, A. Gleason, and David N. Pellow (New York University Press, 2016), 133.

66. Davidson, "Chapter XXIII: The Formosan Tea Industry," 379–380.

67. Thornber, "Afterward," 239.

68. Thornber, "Afterward," 239.

69. Graham Huggan and Helen Tiffin, *Postcolonial Ecocriticism: Literature, Animals, Environment* (Routledge, 2010), 3.

70. Said, *Orientalism*, 3, 6.

BIBLIOGRAPHY

Aghoghovwia, Philip. "Postcolonial Nature." In *Nature and Literary Studies,* edited by Peter Remien and Scott Slovic, 211–228. Cambridge: Cambridge University Press, 2022.

Allen, Herbert J. "Notes of a Journey through Formosa from Tamsui to Taiwanfu." *Proceedings of the Royal Geographical Society of London* 21 (1877): 258–266. https://rdc.reed.edu/i/6dbd3295-77e2-4448-bdd7-e2abda3387e3.

Ana, Jeffrey Santa, Heidi Amin-hong, Rina Garcia Chua, Zhou Xiaojing, eds. *Empire and Environment: Ecological Ruin in the Transpacific*. Ann Arbor, MI: University of Michigan Press, 2022.

Barca, Stefania. "History." In *Keywords: For Environmental Studies*, edited by Joni Adamson, A. Gleason, and David N. Pellow, 132–134. New York: New York University Press, 2016.

Branch, Michael P. "Indexing American Possibilities: The Natural History Writing of Bartram, Wilson, and Audubon." In *The Ecocriticism Reader: Landmarks in Literary Ecology*, edited by Cheryll Glotfelty and Harold Fromm, 282–302. Athens, GA: University of Georgia Press, 1996.

Carrington, George Williams. *Foreigners in Formosa 1841–1874*. Ann Arbor, MI: University of Michigan Press, 1973.

Chang, Chia Ju and Scott Slovic, eds. *Ecocriticism in Taiwan: Identity, Environment, and the Arts.* Lanham, MD: Lexington Books, 2015.

Colquhoun, A. R., and J. H. Stewart-Lockhart. "A Sketch of Formosa." *The China Review* 13 (1885): 161–207. https://rdc.reed.edu/i/2fb0712d-3ff6-4a71-acfc-1fded48fd6d3.

Davidson, J. W. *The Island of Formosa: Past and Present. History, People, and Commercial Prospects. Tea, Camphor, Sugar, Gold, Coal, Sulphur, Economical Plants, and Other Productions.* London and New York: Macmillan & Co., 1903. https://archive.org/details/islandofformosap00daviuoft.

Davidson, J. W. "Chapter XXIII: The Formosan Tea Industry." In *The Island of Formosa: Past and Present. History, People, and Commercial Prospects*, 371–376. London and New York: Macmillan & Co., 1903.

DeLoughrey, Elizabeth, Renee Gosson, and George Handley. *Caribbean Literature and Environment: Between Nature and Culture*. Charlottesville, VA: University of Virginia Press, 2005.

Dodd, John. "Formosa." *The Scottish Geographical Magazine* XI (1895): 553–570. https://rdc.reed.edu/i/784ca96e-71ff-4443-a8f2-752a748a9a62.

Dooren, Thom Van and Deborah Bird Rose. "Lively Ethnography: Storying Animist Worlds." *Environmental Humanities* 8, no. 1 (2016): 77–94.

Estok, Simon and Won-Chung Kim, eds. *East Asian Ecocriticisms: A Critical Reader*. New York: Palgrave Macmillan, 2013.

Grove, Richard H. *Green Imperialism: Colonial Expansion, Tropical Island Edens and the Origins of Environmentalism, 1600–1860.* Cambridge: Cambridge University Press, 1995.

Hancock, William A. "Tamsui Trade Report. For the Year 1881." *China Imperial Maritime Customs. Reports on Trade at the Treaty Ports, For the Year 1881*, 1–38. Statistical Department of the Inspectorate General, 1882. https://rdc.reed.edu/i/07fa4929-760a-4544-84e3-4347bd952875.

Huggan, Graham, and Helen Tiffin. *Postcolonial Ecocriticism: Literature, Animals, Environment*. New York: Routledge, 2010.

Kirksey, S. Eben, and Stefan Helmreich. "The Emergence of Multispecies Ethnography." *Cultural Anthropology* 25, no. 4 (2010): 545–576.

Murphy, Patrick D. *Farther Afield in the Study of Nature-Oriented Literature.* Charlottesville, VA: University Press of Virginia, 2000.

Nichols, Ashton. "Natural History." In *Nature and Literary Studies,* edited by Peter Remien and Scott Slovic, 122–138. Cambridge: Cambridge University Press, 2022.

Otness, Harold M. *One Thousand Westerners in Taiwan, To 1945; A Biographical and Bibliographical Dictionary*. Taipei, Taiwan: Academia Sinica, 1999.

Rose, Deborah Bird. "Ethnography." In *Keywords: For Environmental Studies*, edited by Joni Adamson, William A. Gleason, and David N. Pellow, 110–112. New York: New York University Press, 2016.

Ryan, John Charles. "Beyond East Asian Landscapes: A Survey of Ecocriticism in Thai Literary Studies." In *Referred Proceedings of the International Conference on Trends in Economics, Humanities and Management,* edited by Jiri Strouhal and Michel Plaisent, 26–30. Pattaya, Thailand, 2015. Accessed March 8, 2017, http://icehm.org/upload/3350ED815006.pdf.

Ryan, John Charles. "If We Return, We Will Learn: Empire, Poetry and Biocultural Knowledge in Papua New Guinea." In *Empire and Environment: Ecological Ruin in the Transpacific*, edited by Jeffrey Santa Ana, Heidi Amin-hong, Rina Garcia Chua, and Zhou Xiaojing, 94–110. Ann Arbor, MI: University of Michigan Press, 2022.

Said, Edward. *Orientalism*. New York: Pantheon, 1978.

Scholnick, Robert J. "Permeable Boundaries: Literature and Science in America." In *American Literature and Science*, edited by Robert J. Scholnick, 1–17. Lexington, KY: University Press of Kentucky, 1992.

Steere, Joseph Beal. Letters from Formosa, published in the *Ann Arbor Courier*: LXXI Amoy China 21 Sep 1873 (published in 10 April 1874 *Ann Arbor Courier*);

LXXII Takao Formosa 10 Oct 1873 (published in 10 April 1874 *Ann Arbor Courier*); LXXIII Tamsui Formosa 1 Dec 1873 (published in 24 April 1874 *Ann Arbor Courier*); LXXIV Tamsui Formosa (published in 1 May 1874 *Ann Arbor Courier*); LXXIV [sic] Tamsui Formosa 1 Dec 1873 (published in May 1874 *Ann Arbor Courier*); LXXV Tamsui Formosa 1 Dec 1873 (published in 22 May 1874 *Ann Arbor Courier*); LXXVII Taiwanfu 10 Jan 1874 (published in 5 Jun 1874 *Ann Arbor Courier*); LXXVIII Taiwanfu 1 Feb 1874 (published in 12 Jun 1874 *Ann Arbor Courier*); LXXIX Canton China 10 April (published in 3 July 1874 *Ann Arbor Courier*). https://rdc.reed.edu/i/e8947d6f-a3a5-4439-a321-7e54032ea723.

Steere, Joseph Beal. *Formosa and Its Inhabitants*. Edited by Paul Jen-Kuei Li. Taipei: Institute of Taiwan History (Preparatory Office), Academia Sinica, 2002.

Steere, Joseph Beal. "Formosa." *Journal of the American Geographical Society of New York* 6 (1874): 302–34. https://rdc.reed.edu/i/4cd68be9-1612-45fb-b739-95012d7b80e7.

Sweet, Timothy. "Economy, Ecology, and Utopia in Early Colonial Promotional Literature." *American Literature* 71, no. 3 (1999): 399–427.

Swinhoe, Robert. "Notes on the Island of Formosa." *Journal of the Royal Geographical Society of London* 34 (1864): 6–18. https://rdc.reed.edu/i/2fee8170-ee11-4de6-b5d4-fc09a4cf36c6.

Tallmadge, John. "Toward a Natural History of Reading." In *The ISLE Reader: Ecocriticism, 1993–2003*, edited by Michael P. Branch and Scott Slovic, 282–295. Athens, GA: University of Georgia Press, 2003.

Telles, Jason Paolo, John Charles Ryan, and Jeconiah Louis Dreisbach, eds. *Environment, Media, and Popular Culture in Southeast Asia*. Singapore, Springer, 2022.

Thornber, Karen Laura. *Ecoambiguity: East Asian Literatures and Environmental Crises*. Ann Arbor, MI: University of Michigan Press, 2015.

Thornber, Karen Laura. "Afterward: Ecocritical and Literary Futures." In *East Asian Ecocriticism: A Critical Reade*r, edited by Simon C. Estok and Won-Chung Kim, 237–258. New York: Palgrave Macmillan, 2015.

Welch, Margaret. *The Book of Nature: Natural History in the United States 1825–1875.* Boston: Northeastern University Press, 1998.

Worster, Donald. *Nature's Economy: A History of Ecological Ideas*. New York: Cambridge University Press, 1994.

Chapter 3

Oriental Beauty

The Tea Plantations and Global Tea Trade in Nineteenth-Century Taiwan

Iping Liang

Anyone who visits Taiwan today will surely be introduced to the most iconic drink—the Bubble Tea. Made of tapioca balls, brown sugar, milk, and tea, it is also called the "pearl milk tea." The tea is sweetened by the chewy and queey[1] "pearls" that are made of tapioca balls to be topped by red beans, green beans, barleys, soybean pudding, or anything else that your heart desires for. Taiwanese Bubble Tea has been globalized in the last twenty years: it is now available in Berlin, London, Los Angeles, Shanghai, Singapore, Jakarta, and Ho Chin Ming City. When Hilary Clinton was running for the presidential campaign in 2016, she tried it out in Flushing, New York. Sipping the bubbles, she cried out, "Oh! I love it. I've never had chewy tea."[2] Undoubtedly, the Bubble Tea has become one of the most popular and trend-setting commodities from Taiwan.

The Bubble Tea, however, is not the first globalized tea product from Taiwan. In the late nineteenth century, the Formosa Oolong Tea had been a phenomenal success. It was so popular that it was showcased in a specially designed Taiwan Tea House at the St. Louis World Fair in 1904.[3] The two success stories, although about 150 years apart, testify to the popularity of Taiwanese tea commodities in the global arena. This chapter will examine the representation of the tea plantations in Taiwanese woman writer Jade Yu-Huei Chen's novel, *The Merry Leaf* (2014), in the context of the Plantationocene. By drawing on the work of anthropologist Anna L. Tsing (2015), biologist Donna Haraway (2019), environmental philosopher Bruno Latour (2020), etc, it unravels the rise of Formosa tea trade in the 1860s against the background of imperial botanical exploration that was initiated by British plant hunters such as Robert Fortune (1812–1880), Henry John Elwes (1830–1922), and most importantly by British tea merchant John Dodd (1838–1907). It argues that the commercialization of Formosa Oolong Tea and the branding of Oriental Beauty, which was presumably named by Queen Elizabeth,[4] discloses the inseparable relation between British imperialism and global tea trade.

Divided into four parts, this chapter will examine the cultivation of the tea plantations in northern Taiwan in the nineteenth century and the literary representation of the global tea trade in the novel. The first part will provide a general introduction. The second section will examine the rise of the global tea trade in the context of British botanical imperialism in the second half of the nineteenth century. The third will analyze the literary representation of the tea plantations in the context of the Plantationocene. The fourth part will make the conclusion.

BRITISH IMPERIALISM AND THE GLOBAL TEA TRADE

Tea making is a labor-intensive and time-consuming process. It may not be as "merry" as John Dodd once described.[5] This section will focus on the close relationship between plant and empire and examine how the success of Formosa Oolong Tea was closely related to nineteenth-century British imperialism. I will start with Linda Schiebinger and her work on "plant mercantilism" and examine the adventure of plant hunter Robert Fortune in the context of British imperialism. I will then relate Fortune with his follower John Dodd, who was most instrumental in the branding of Formosa Oolong Tea. I will then investigate the relationship between the global tea trade and the notion of the Plantationocene, which was conceptionalized by Anna Tsing and Donna Haraway in 2015.

The work of Londa Schiebinger has been instrumental to the investigation of the relationship between the imperialist foundation and botanical explorations. In *Plants and Empire: Colonial Bioprospecting in the Atlantic World* (2004), she argues, "historians of the eighteenth century have begun to detail how botany—expertise in bioprospecting, plant identification, transport, and acclimatization—worked hand-in-hand with European colonial expansion."[6] While the establishment of European botanical gardens since the 1500s had paved the road for imperial and colonial exploration of various plants—especially spice, sugarcane, ginger, and breadfruit,[7] Amitav Ghosh in *The Nutmeg's Curse: Parables for a Planet in Crisis* (2021) comments on how the "monopoly of nutmeg and mace" on Banda Island by the VOC in the early 1600s could be related to the extinction of the Pequot in the New

World around the same time. He says, "In both [cases,] large numbers of captives were enslaved and transported overseas to work on plantations; and both were intended to extinguish a people."[8] That the practice of native genocide, especially in the case of the Bandanese, was related to a particular species of vegetation is certainly alarming. Yet, this is exactly the point that Schiebinger has contended: botanists (or naturalists) were "agents of empire."[9] Their training in "inventories, classifications, and transplantations were the vanguard and 'instruments' of European order."[10] Bruno Latour also observes that "the technologies of collection—both material and intellectual—had extended the imperial power of European nations."[11] That is to say, as a discipline of science, the rise of botany in the late eighteenth century testified to the expansion of European colonial power—a point Jeffery Nealon also makes in *Plant Theory: Biopower and Vegetable Life* (2015).[12]

It is in the context of "colonial bioprospecting," or "plant mercantilism," that I will turn to the life and career of plant hunter Robert Fortune. In her book, *For All the Tea in China: How England Stole the World's Favorite Drink and Changed History* (2010), Sarah Rose notes that in the wake of the Opium War, England realized the need to control the access to tea. Governor-General Henry Hardinge claimed, "In my opinion, the cultivation of tea is likely to prove an equally prolific and more safe source of revenue than that derived from the monopoly of opium."[13] To do that, England would need "healthy specimens of finest tea plants, seeds by the thousand, and the task required a plant hunter, a gardener, a thief, and a spy."[14] That man was Robert Fortune. Born in Berwickshire, Scotland in 1812, Fortune acquired hands-on knowledge about plants and flora from his father, who was a respected gardener. He grew up to become a well-known botanist and gardener and was later hired by the Kew Garden to bring home the seeds of tea trees as a strategic resource for British imperial operation. He arrived in China in 1842 and stayed there for three years. By 1845, Fortune had shipped back to England the seeds of tea plants that would later be harvested in Assam and Darjeeling in great quantities, thus disrupting the monopoly of tea production in China. Fortune was later considered as the "Father of tea industry in England."

While Fortune had indeed materialized the big fortunes for the British empire, the massive cultivation of tea plantations had also caused serious damage to the local environment. In her dissertation, "Brewing Trouble or Transforming Nature? The Making of Tea Plantation Environments in Assam, 1830s–1930s" (2021), Namrata Borkotoky contends,

> Tea plantations in Assam brought about significant modifications in the physical environment along with the well-documented socio-economic and political transformations. . . . The emergence of tea as an industry in Assam coincided with the simultaneous initiation, growth and influx of intellectual or scientific

> discourses which included agricultural expertise, practical "on-field" expertise, and various scientific and technological researches. Built around the narrative of tea cultivation prevalent among this colonial tea "expertise," who implicitly dictated how the environment in these plantations would be manipulated, the thesis explores a common thread that binds the themes under study—deforestation, soil degradation, pest occurrences and the corporeal degeneration of its laborers.[15]

As Borkotoky observes, the "manipulation" of tea plantations by colonial tea experts resulted in "deforestation, soil degradation, pest occurrences and corporeal degeneration of [tea] laborers." The practice seems to "brew trouble" and "transform nature" at the same time. Placing Assam tea plantations in the context of environmental history, she aims to "explore the inter-connectedness between human actions in the tea environs and its unforeseen and long-lasting consequences in the plantation environments."[16] Likewise, David Montgomery argues that European empires had introduced the large-scale single-crop cultivation, such as sugar, coffee, tea, and pine apple, that had severely reduced biodiversity and soil fertility of the colonized regions.[17] Moreover, Sarah Rose points out the fact of colonial encroachment by making an assessment of the impact of tea plantations on the local environment in India—

> Today, we have guarded enthusiasm for the mass globalization of Indigenous plant life. We know now that when species are brought to new habitats, they can overpopulate and decimate local eco-systems. Entire islands have been overrun as the result of the kind of botanical frontiersmanship that [Robert] Fortune and his contemporaries had routinely practiced.[18]

Taken together, eco-critics and environmental historians such as Borkotoky, Montgomery, and Rose have assessed the negative impact the tea plantations had brought about to both China and India. The collective damage in deforestation, soil degradation, pest occurrences, labor degeneration, and loss of biodiversity, among others, resulted in the drastic transformation of the local environment.

I further argue that the practice of "botanical frontiersmanship" in the tea plantations, as Rose points out, could be considered as an assault of the Plantationocene. First conceptualized by Anna Tsing and Donna Haraway in 2015, the idea of the Plantationocene was initially brought up in a roundtable they took part in Denmark. In their then-co-authored essay, "Anthropologists Are Talking about the Anthropocene," they were concerned with the issue of the long history of slave agriculture as a key transition in the plantations in the Caribbean and Southeast. They pointed out that the colonial plantation system made possible a large-scale "simplification of landscapes all over the world, [as well as] the transportation of genomes, of breeding plants, animals, and the alienation between the crops and the farmers" (556). In the context of colonialism, they observed that "we need to understand the dynamics through which plants and animals are abstracted in order to become resources that can be used for investment. Plantations and feedlots are places where this happened" (556). Japanese anthropologist Noboru Ishikawa added that "plantations are just the slavery of plants" (556). Moreover, geographer Kenneth Olwig argued that the plantation was also related to a manipulative scheme of the landscape in terms of a large-scape cultivation of a single crop. He says,

> It is a space, which is uniform in its own economic context, but does not really exist in the greater scheme of earthly life. The next step is to try to manipulate this new reality, through drainage so that you get crops that grow evenly in spatially uniform square of property. You are manipulating the world to make it fit a utopian Euclidian grid that maybe is good to think with and good to construct property relation with. (558)

This "Euclidocene" was particularly regulated by the spatially large-scale cultivation of a single crop. The uniformity in the landscape had been observed in the tea plantations in both India and China. Thus, in their subsequent essay, "Reflections on the Plantations: A Conversation with Donna Haraway and Anna Tsing" (2019), Haraway explicates that three things seem to have dominated the plantation for the past 500 years—"radical simplification; substitution of peoples, crops, microbes, and life forms; forced labor; and, crucially, the disordering of times of generation across species, including human beings" (6). Tsing adds by saying that "the plantation was precisely the conjuncture between ecological simplifications, the discipline of plants in particular, and the discipline of humans to work on them" (6). In other words, the Plantationocene, according to Haraway and Tsing, is the reason for forced or coerced labor, displacement, and dispossession (10–11) in the large-scale production of a single crop as a way for "plant mercantilism"—the integration of the tea plantations and the global tea trade. In the following, I will turn to the literary representations of the tea plantations and the global tea trade in *The Merry Leaf*.

THE REPRESENTATION OF THE TEA PLANTATIONS IN *THE MERRY LEAF*

Taking my cue from Tsing, Haraway, Rose, and so forth, I turn to the representation of the tea plantations in *The Merry Leaf* in the context of the Plantationocene. I will first introduce the author and the reception of the novel before I examine the literary portrayal of the tea plantations and the global tea trade in the novel. Jade Y. H. Chen is a well-established writer from Taiwan. Born in Taichung (in central Taiwan), she went abroad in her youth and studied French, acting, and performance arts in France and Germany. Moreover, she has been an active journalist stationed in Europe for many years. Married to a German man, she divides her time between Germany, other parts of Europe and Taiwan. To date, she has published 11 screenplays, 12 novels, and 18 collections of prose. Her novels have won major awards in Taiwan.[19] *The Merry Leaf* is one of her books that are related to material culture, such as china and tea. On the back cover of the book, the blub says, "In the middle of the 19th century, a Chinese woman traveled alone to Taiwan, getting to know the Scottish explorer John Dodd and Taiwanese merchant Chun-sheng Li. Against the background of the global tea trade, *The Merry Leaf* tells the most romantic story in the history of tea plantations in Taiwan."[20] Set in the 1860s, the novel is a combination of both facts and fabrication. The story centers on a legendary female tea maker, Zhi-yun Wei, and the three men in her life. First, it is John Dodd, who was the famous Scottish/British tea merchant who came to Taiwan in 1860 to start the tea trade in Dadaocheng, the oldest commercial district of Taipei. Second, it is Chun-sheng Li, the Chinese comprador who helped Dodd establish the tea business in northern Taiwan. Li is special due to his ability to speak English: he was baptized and learned English in church. The third character is an imaginary Chinese tea maker, Chi-hua Kao, who is a foster son of Wei's family. Chi-hua grows up with Zhi-yun and has secretly loved her for many years. The novel highlights the successful branding of Formosa Oolong Tea, featuring Zhi-yun as the "Oriental Beauty" and John Dodd and Chun-sheng Li as the "Fathers of Taiwanese Oolong Tea." The success of the tea trade however does not ease the rivalry between John Dodd and Chi-hua Kao, as both are in love with Zhi-yun. The novel ends with Chi-hua taking his son back to China, John Dodd going back to England sadly, while Zhi-yun dies of the complication of childbirth.

While the novel does not seem to tell a "merry" story, it has received many positive reviews. Most critics agree that Chen tells a love story to draw the reader's attention to the history of Formosa Ooloong Tea, which seems to have been forgotten.[21] The famous critic Fang-ming Chen pays attention to the theme of diaspora in the novel. He argues that the fact that Zhi-yun comes from China and dies on the island of Taiwan seems to suggest a diasporic journey and the nativist resettlement of Zhi-yun. That is to say, Zhi-yun symbolizes nativism in Taiwan, making the island her final home.[22] While Chen highlights the aspect of national allegory, Ning-Hsin Yang focuses on that of material culture. In her thesis, "A Study on Jade Chen's *China* and *The Merry Leaf*: Space and Object Perspectives" (2015), Yang explores spatial identities and material symbolism in relation to the tea trade. She contends that "the transcultural reading of the tea trade and the spatial travels helps us understand the embedded cultural identities of the novel."[23] Finally, Hue-lan Huang's thesis, "A Study of the Historical Types of Taiwan's Commercial War Novels" (2019), focuses on the narrative type of the "commercial war" and teases out the steep competition of the tea trade during this period.[24] Huang argues that "*The Merry Leaf* takes the global tea trade as the historical background, thus reconstructing the economic context of Taiwan."[25] She analyzes Chen's creative writing, as well as the three topics of the novel, including "the reproduction of history," "commercial war writing," and "material circulation and spatial alteration." Taken together, it is interesting to note that critics are well-balanced in their assessment of the ethos, aesthetics and geopolitics of the novel.

While retaining my interest in the Plantationocene, I will develop my analysis of the novel in the following three aspects. First, I argue that John Dodd is a successor of Robert Fortune in the context of Formosa Oolong Tea. Like Fortune, Dodd is interested in "plant mercantilism" that Londa Schiebinger has defined. The successful branding of Formosa Oolong Tea is a remarkable example of "plant mercantilism." Second, I maintain that Dodd's search for the tea seedlings and a fitting place for tea cultivation could be considered as a case of "colonial bioprospecting"—the expertise "in plant identification, transport, and acclimatization" and how it "worked hand-in-hand with European colonial expansion" as Schiebinger has suggested. Third, I content that the practice of plant mercantilism is based on coerced labor and profit, which results in not only the alienation of the planter from the crop, but also the final betrayal of Chi-hua to Zhi-yun. Below I will elaborate on these points.

First, I argue that John Dodd is a successor of Robert Fortune in the case of Formosa Oolong Tea. Like Fortune, he is interested in "plant mercantilism" and carries out "colonial bioprospecting" that Schiebinger has described. At the beginning of the novel, Dodd is introduced as a typical Western explorer who is fascinated by Asia. Arriving in Taiwan in 1860, he joined naturalists, explorers and diplomats such as William Pickering (1840–1903), Robert Swinhoe (1836–1877), Charles le Gendre (1830–1899),[26] etc. to explore the foreign island. In the novel, he is portrayed as—

> Coming from Westmorland, Scotland, Dodd's childhood dream was to explore in Asia, especially China and India. He has been greatly inspired by the stories of the geographical discovery. He was only twenty-two years when he boarded the ship of the British East India Company. Starting as an apprentice, he had spent five years on the ship, listening to tall tales of the captain and fellow seamen."[27]

At the juncture of the story, Dodd has arrived in Hong Kong and will soon start to work as a clerk for Mr. Spenser at the Elles & Co. One thing peculiar about Dodd is his disgust against opium. His father was addicted to opium and died alone on a cold winter day.[28] Dodd is enthusiastic about the prospect of Asia. Seeing his "vision," he is eager to make his dream come true. Looking ahead, he says, "The Suez Canal is now already open and China has to open five trading ports. Our opportunities are final here."[29] He soon befriends Chun-sheng Li, "a self-righteous man, well-behaved, very smart and quick-witted."[30] Dodd and Li become instant partners and make the decision that they will look into the prospect of tea cultivation in Taiwan. Moreover, Dodd's interest in tea may be influenced by Fortune as it is mentioned in the novel that Dodd has been greatly inspired by the career of Fortune. He has read the book about Fortune's adventures in China and how he had used the Wardian cases to ship the tea seedlings out of China.[31] Knowing that Fortune had shipped the seedlings to Darjeeling and planted hundreds of tea trees there, Dodd is, instead, interested in the cultivation of green tea. In addition, Dodd's interest in tea is greatly encouraged by Robert Swinhoe, who in the novel informs him that he has seen tea trees in Nantou[32] and that it had been recorded by the Dutch in 1645 that "the soil of Taiwan is fitting for tea cultivation. There have been wild tea plants."[33] Their search for wild tea plants is also confirmed by an elder who treats them a cup of good tea and takes them to see the wild tee plant. Dodd is thus convicted that Taiwan would be a fitting place for tea cultivation.[34] He then joins Li to visit Zhi-yun in Anxi, Fukien and asks her to help them plant teas for their plantations in northern Taiwan. Li is in charge of the shipping of 3,000 tea seedlings from Fukien to Taiwan, by way of Amoy.[35] The process of surveys and field trips would fit the notion of "colonial bio-prospecting" that Schiebinger has described—the "expertise in bioprospecting, plant identification, transport, and acclimatization"—which also works hand-in-hand with the expansion of tea plantations in northern Taiwan as both Dodd and Li are passionately engaged in.

Moreover, I argue that Dodd and Li collaborate in their commercial scheme to convert local farmers to single-crop tea cultivation. After they have successfully shipped the tea seedlings to Tamsui in northern Taiwan, Dodd and Li make trips to Sanxia to recruit more tea planters. Traditionally, the local farmers in Sanxia have planted potatoes, barleys, bennes, sesames, and indigos. The area has been known for its indigo dye. Since the discovery of artificially-synthetic indigo, the sale has dropped drastically. As a result, the farmers are forced to turn to other crops. Knowing that they are mostly from Anxi, Dodd invites them to consider a tea plantation—what Anxi is known for historically. Li adds by saying, "We will provide you with tea seedlings, free of charge. You just plant them. We will buy the harvest when it is ready. As long as you have tea leaves, we will buy them. What do you think?"[36] This wholesale package of free seedlings, high acquisition price, and the provision of set-up loans is so attractive that it soon proves to be irresistible. By doing so, Dodd and Li are able to convert farmers in the rural areas of Taipei, such as Muza, Shenkeng, and Shiding, to the cultivation of tea plantations, thus resulting in unprecedented large-scale single-crop agriculture in northern Taiwan. In addition, the conversion to tea plantations has become so pervasive that it is an encroachment to the traditional homelands of the Indigenous people. As the forests have been cleared and the tea plantations have sprawled to high mountains, the Indigenous people feel offended and threatened. As a result, they conduct a traditional headhunting ceremony and behead a local farmer to express their anger and protest. Dodd and Li end up making a trip to the mountains to meet the chief and negotiate a mutual agreement for peace and co-existence.[37]

Finally, the rise of Formosa Oolong Tea also results in the competition between tea traders in terms of commercial warfare. An interesting plot twist is that while Zhi-yun creates fame for herself in association with Formosa Oolong Tea, her childhood lover Chi-hua does not approve of her "innovation." Chi-hua does not like the light flavor of Formosa Oolong Tea and is never comfortable with the change. He then becomes the target of a fellow tea trader, Ping-yuan Wang, who is a keen contestant of the Wei family back in Anxi, Fukien. Knowing that Chi-hua is also a skillful tea master, Wang approaches Chi-hua secretly and offers him a marriage proposal to his sister, Li-zhi. Wang tells Chi-hua that his sister likes him so much that she would not consider anyone else for marriage. Being rejected by Chi-hua, Wang then offers him the right to plant the tea in Wang's family, where he can cultivate the tea by following the traditional way. Still being turned down, Wang then raises the stakes and offers the ownership of the Wang family's entire tea plantations if Chi-hua agrees to marry his sister. That is to say, Wang would give Chi-hua the right of inheritance to his family tea plantations which are far bigger than those of the Weis. As Chi-hua gradually feels unsure and ambivalent toward Zhi-yun—his fundamental disagreement with her in the novelty of Formosa Oolong Tea, and his total disgust against John Dodd, who is a foreigner after all—he finally caves in. While we could understand the nature of commercial war between the tea traders, it is needless to say that Chi-hua is also sexually attracted to Li-zhi, who agrees to his every bodily aggression.[38]

CONCLUSION

Knowing the secret affair between Chi-hua and Li-zhi, Zhi-yun then yields in and agrees to marry John Dodd. Showing his love for her, Dodd paints the portrait of Zhi-yun on the tin boxes of the tea. Formosa Oolong Tea then makes a hit in New York and joins the global tea trade. According to Man-hong Lin, whose *Industries of Tea, Sugar, and Camphor and Socio-Economic Change of Taiwan, 1860–1895* has been a classic in the field, "Between 1871 and 1896, the export of Taiwanese tea grew 12 times."[39] Chen likewise writes in the novel, "in 1885, the tea export from Taiwan is six times more than that from Xiamen."[40] Shortly after the marriage, Dodd receives a letter from Queen Elizabeth, who writes to him and thanks him for his time and efforts in the making of Formosa Oolong Tea. In appreciation of his contribution, Her Majesty honors him with a royal name for the tea "Oriental Beauty."[41]

In conclusion, Chen's characterization of Zhi-yun as Oriental Beauty is well done. The lure of the "tea fairy" in the novel is beyond doubt. While the imaginary love story adds much flavor to the history of Formosa Oolong Tea, I also argue that the development of the tea plantations in northern Taiwan in the late nineteenth century should be considered as an example of the Plantationocene as this chapter has articulated. The contribution of historical figures such as John Dodd and Chun-sheng Li should be recognized in terms of agricultural economy and the development of global tea trade. As a result, Taiwan was re-incorporated into the network of global trade in the late nineteenth century, which made the island more profitable in the eye of Japan. Thirty-five years after Dodd arrived in Taiwan, the island was colonized by Japan in 1895. While the issue of "colonial bioprospecting" in association with figures such as John Dodd, Robert Swinhoe, Charles le Gendre, and so on, deserves further investigation, I conclude that their participation in and assistance with "botanical frontiersmanship" should be historically understood in the context of the Plantationocene.

NOTES

1. This is a local vernacular term that has become popular with the bubble tea. It refers to the stickiness of tapioca—the flavor that makes the bubble tea special.

2. Please see https://www.vice.com/en/article/gvk3xb/hillary-clinton-tried-boba-for-the-first-time-and-called-it-chewy-teathe newspaper article.

3. Please see Hsin-yi Lin (2023).

4. This has not been authenticated but widely circulated among the circles of local tea communities in Taiwan. For details, please check https://www.jsy-tea.com/blog/posts/oriental-beauty.

5. According to Chen, the title of the novel, "the merry leaf," was inspired by a comment made by Dodd, when he saw tea-picking women workers were "merrily" chatting among one another in the tea workshop.

6. Schiebinger, 7.
7. Schiebinger, 4.
8. Ghosh, 21.
9. Schiebinger, 11.
10. Ibid.
11. Qtd. in Schiebinger, 11.
12. Nealon argues that in the wake of natural history, vegetation was not considered part the realm of "life." He reads Foucault's *The Order of Things*—"The naturalist is the man concerned with the structure of the visible world and its denomination according to characters. Not with life" (*OT* 161), Foucault insists, because life is not representable. Life is in fact a kind of unplumbable depth, animating the organism from a hidden origin somewhere within. This birth of biology—which is to say, the emergence of 'life' itself as a bearing area for discursive power and a depth to be explored—constitutes the first birth of biopower," which by definition excluded the world of plants (36).
13. *For All the Tea*, 8.
14. Ibid., 12.
15. For details, please see https://shodhganga.inflibnet.ac.in/handle/10603/420594.
16. Ibid., 2.
17. Please see Montgomery, p. 110.
18. Ibid., 268.
19. For example, an earlier novel, *Mazu's Bodyguards* (2004), was granted the Golden Award of Taiwan Literary Prize (2004) and the Jury's Prize of the Red Chamber Dream in the Baptist University of Hong Kong (2005), etc.
20. This is seen in the back cover of the book, which offers a plot summary to help boost the sale of the novel.
21. Please see the interview with Chen.
22. Chen's criticism is found in the back cover of the novel.
23. For details, please see the link, https://www.eslite.com/product/10012233523 05419.
24. For details, please see the link, https://hdl.handle.net/11296/ncfq67.
25. For details, please see the link, https://hdl.handle.net/11296/457m93.
26. Pickering, Swinhoe, McKay and Le Gendre all appear in the novel. Pickering and Swinhoe provide help and advice for tea cultivation; McKay hosts the wedding for Dodd and Zhi-yun while Le Gendre appears at one of the parties.
27. *The Merry Leaf*, 13.
28. Ibid., 15.
29. Ibid., 16.
30. Ibid., 18.
31. Ibid., 23.
32. Ibid., 28.
33. Ibid., 29.
34. Ibid., 29.
35. Ibid., 57.
36. Ibid., 73.

37. Ibid., 76–78.
38. Ibid., 288–304.
39. Lin, 19.
40. Chen, 306.
41. Chen, 340.

BIBLIOGRAPHY

Beattie, James, Edward Melillo, and Emily O'Gorman. "Rethinking the British Empire through Eco-cultural Networks: Materialist-cultural Environmental History, Relational Connections and Agency." *Environment and History* 20, no. 4 (2014): 561–575. https://www.jstor.org/stable/43299706 [MSU]

Borkotoky, Namrata. "Brewing Trouble or Transforming Nature? Making of Tea Plantation Environments in Assam, 1830s–1930s." PhD diss., Indian Institute of Technology Guwahati, 2021. http://gyan.iitg.ernet.in/handle/123456789/2046

Chen, Jade Y. H., *The Merry Leaf*. Taipei: Ink, 2014.

Dey, Arnab. "Bugs in the Garden: Tea Plantations and Environmental Constraints in Eastern India (Assam), 1840–1910." *Environment and History* 21, no. 4 (2015): 537–565. https://www.jstor.org/stable/24810625 [MSU]

Etherington, Dan M., and Keith Forster. "The Structural Transformation of Taiwan's Tea Industry." *World Development* 20, no. 3 (1992): 401–422. https://doi.org/10.1016/0305-750X(92)90032-Q [MSU]

Gardella, Robert. *Harvesting Mountains: Fujian and the China Tea Trade, 1757–1937.* Berkeley: University of California Press, 1994.

Huang, Huei-lan, "A Historical Study of Taiwanese Commercial War Fiction," Thesis, National Taiwan University of Education, 2019.

Lin, Hsin-yi. "Caught between Japan and the US: Taiwan's Participation in the 1904 St. Louis World Fair." *NTNU Journal of Historical Studies* 69 (2023): 111–166. DOI:10.6243/BHR.202306_(69).0004

Lin, M. H. *The Industries of Tea, Sugar, and Camphor and Socio-Economic Change of Taiwan, 1860–1895* (*Cha tang zhangnao ye yu Taiwan zhi sheshui jingji bianqian, 1860–1895*). Taipei: Lien-Ching, 1997.

Lu, Li-Ru. "Ecocriticism of Nineteenth-Century Taiwan in John Dodd's 'Formosa.'" *The International Journal of Critical Cultural Studies* 15, no. 2 (2017): 1–8. https://doi.org/10.18848/2327-0055/CGP/v15i02/1-8

Montgomery, David. *Dirt: The Erosion of Civilization*. Berkeley: The University of California Press, 2007.

Nealon, Jeffrey. *Plant Theory: Biopower and Vegetable Life*. Stanford: Stanford University Press, 2015.

Rose, Sarah. *For All the Tea in China: How England Stole the World's Favorite Drink and Changed History*. New York: Penguin, 2010.

Ru, Sung Hee. "Conceptualizing China's Tea History in the 19th Century: Incorporation into the Capitalist World□Economy." *Journal of Agrarian Change* 22, no. 4 (2022): 812–830.

Sarkar, Tahiti. "Understanding the Transformation of Colonial Darjeeling Hills through the Study of South Asian Environmental History." *NBU Journal of History* 9 (March 2016): 223–231. https://ir.nbu.ac.in/handle/123456789/3882

Sarkar, Tahiti. "Linking Environmental History of South Asia with Colonial Darjeeling Hills: A Unique Historical Palimpsest." *Salesian Journal of Humanities and Social Sciences*, 9, no. 1 (2018): 80–89. https://ssrn.com/abstract=3642468

Schiebinger, Londa, *Plants and Empire: Colonial Bioprospecting in the Atlantic World.* Cambridge: Harvard University Press, 2004.

Tsing, Anna and Donna Haraway, "Reflections on the Plantations: A Conversation with Donna Haraway and Anna Tsing," *Edge Effects* (2019)

Yang, Ning-Hsin. "A Study on Jade Chen's *China* and *The Merry Leaf*: Space and Object Perspectives," master's thesis, National Taiwan Normal University, 2015.

Chapter 4

From Resonance to Solastalgia

Tea Poems of Taiwan, ca. 1820–1920

Stephen Roddy

For over a millennium (ca. 800–1900), the genre of short lyric poetry known as *zhuzhici* ("bamboo branch lyrics") served as a vehicle for conveying psychosocial interactions of human communities with local geographical phenomena. During the final outpouring of this poetry in the eighteenth through the early twentieth centuries, they often allude to increasingly unprecedented socioeconomic and environmental conditions that were reshaping the Sinographic world, including Japan and Korea as well as China itself. This was especially true in Taiwan, where demographic and environmental changes accelerated both before and after Japan annexed the island in 1895. Though mostly written by classically-educated literati, such poems typically center on the experiences of farmers, miners, fishermen, or other laborers—that is, populations who experienced ecological and demographic volatility at a visceral level. Among the many humble occupations treated in this poetry, the work of cultivating, harvesting, and processing *camelia sinensis*, the tea plant, is especially noteworthy.

In the following chapter, I explore Taiwan *zhuzhici*[1] about tea as precursors to contemporary ethnographies that treat both interhuman and interspecies relationships in lyrically evocative forms. In recent decades, ethnographers working in various disciplines have advocated for an affective "attunement" with their subjects that result in "performative texts" invoking "tacit, implicit, and embodied" forms of knowledge.[2] Closely paralleling such embodiment within and through the medium of texts is the call to emphasize the (human) body's "resonance" with materiality, both animate and inanimate, in geographic spaces.[3] Some scholars have likened resonance quite literally to sonority, what some call an in-between-ness that mediates between and among multiple states, encompassing the flora, fauna, geology, and topography, inter alia, of landscapes.[4]

Needless to say, poetic traditions in China and elsewhere are well-suited to articulating a verbal "attunement" or "resonance" between humanity and the varied constituents of surrounding environments. Poetry does so in part through prosodic and broadly literary features that "share a cultural function with myth, performance, and ritual, promoting ways of being that are not instructional but, instead, are performative, relying on the enactment and re-enactment of the verse itself, rather than on argumentation."[5] It is this possibility for re-enactment, a "refrain" that replaces the rigidly empirical modes of academic prose with affectively rich modes of discourse.[6] Drawing inspiration from Félix Guattari and Gilles Deleuze's notion of geophilosophy, such reconfiguring of ethnography has begun to make room for the poet, or more generally, the lyrical ethnographer, attuned to "polyphonic, polyrhythmic surround" in "the world at the local level." In other words, writing that conveys an affectively productive resonance with the environment, broadly defined.

Tea is a particularly fruitful subject for exploring bodily "resonance" with other constituents, sentient or not, of a given locality. A horticultural product that entails varietal selection, pruning, irrigation, and harvesting, it is also subject to different types and styles of roasting or curing, and then prepared in aesthetically and ritually significant forms to be imbibed by—that is, to bring into, the body. What were the forms of labor that contributed to this end product, and who were involved? How does its presence alter or otherwise affect the landscapes that sustain it? What is its relation to poems that center on other agricultural products and natural resources drawn from the same landscapes? And finally, how do some *zhuzhici* presage resonance in a specifically melancholic register: that is, what Glen Albrecht has coined solastalgia? This essay attempts to answer such questions by reviewing representative examples from the corpus of tea poems in the early modern era.

The subject of tea assumed prominence in the Chinese cultural imaginary over a millennium ago, during the mid-Tang Dynasty (late-eighth century CE)—that is, nearly simultaneously with the birth of *zhuzhici*. It was then that the beverage was embraced by literati as both an ethos of elegant deportment and a vehicle for achieving spiritual and cultural profundity. The poet Lu Tong (790–835) and the recluse Lu Yu (733–804, the author of *Chajing*, ["Classic of Tea"]) inspired future generations to pursue multiple aspects of tea connoisseurship, including gustatory, literary, and aesthetic complements to its consumption. Eventually, some poets began to write about the tasks, performed mostly though not entirely by women, of caring for and harvesting tea leaves, often under the title, "tea-picking poems" (*caichashi*). An early example of this is the late-Yuan and early-Ming poet Gao Qi's (1336–1373) "tea-picking song" (which is not a *zhuzhici*) about the tea grown near his home in Suzhou:

> Thunder passes through Xishan, its azure clouds warm
> Poking out halfway from thick bushes, "spear furls" are short
> Girls with silver hairpins answer each other in song
> Seeing who can pick the most for her basket
> Returning home, the clear fragrance lingers on their hands
> They first present the best leaves to the magistrate
> Having just cured them in their wicker stoves, they couldn't taste them
> Filling their baskets, they sell to merchants from Hunan
> Households in the mountains do not know how to plant grain
> Their livelihood depends from year to year on the spring rains
>
> Gao Qi, "Tea Picking Song"

> 雷過溪山碧雲暖　　幽叢半吐槍旗短
> 銀釵女兒相應歌　　筐中摘得誰最多
> 歸來清香猶在手　　高品先將呈太守
> 竹爐新焙未得嘗　　籠盛販與湖南商
> 山家不解種禾黍　　衣食年年在春雨
>
> 高啟,《采茶詞》[7]

The tea fields of Mt. Lu, in the tea-producing region of Wuling in Jiangxi Province, also inspired many later poets, such as in the following quatrain by an early Qing official serving as a magistrate there:

> Returning from picking tea atop Mt. Lu,
> Our clothes completely soaked as we wandered in the clouds and fog
> We've learned new curing methods from Beiyuan [in Zhejiang]
> Tender spikes of Jiangnan tea grow heavy before the rains begin
>
> Jiang Gao (1635–1715), "Middle Yangzi *zhuzhici*"

> 匡廬山上採茶歸　　雲霧迷空盡濕衣
> 學得北源新焙法　　江南嫩甲雨前肥
>
> 江皋,《江中竹枝詞》[8]

Jiang's poem takes the form of a seven-character quatrain (the predominant but by no means only form of *zhuzhici*), and alludes to both processing techniques as well as climatic conditions conducive to producing delectable foliage. This dual focus is characteristic of *zhuzhici* generally, and particularly during the Qing: they describe both "soil and customs" (*tusu*), nature as well as culture, of the localities in which they are set. As in Jiang's poem, many mention local topography such as mountains, rivers, and lakes, along with flora and fauna, weather, or other aspects of the locality. They were sometimes written or read as a kind of shorthand notation that would supplement local gazetteers, in which some were included and thereby preserved for posterity.[9]

From the earliest examples by Liu Yuxi (772–842) and Bai Juyi (772–846), central to the conception of *zhuzhici* were the tasks of labor: sericulture, rice planting and harvesting, fishing, and rowing (a significant proportion are titled *zhaoge*, rowing songs), but also entertainment (by courtesans), preparation of food and drink, mercantile itinerancy, or other humble or mundane activities.[10] In the tea-growing regions south of the Yangzi River, the work of raising, collecting, and producing tea—whether explicitly by women or not—was a common theme, such as in the example by Jiang Gao above. When tea plantations began to expand into the hilly areas of northern and central Taiwan in the early decades of the nineteenth century, a visitor from Jiangxi Province wrote several poems about the tea-picking women of these areas.

> Her ten delicate, slender fingers pluck the new tealeaves
> Her head tented by a wide hat, her face shielded by gauze
> Though a fiery sun beats down from on her crimson parasol, she rests not
> Beads of fragrant perspiration scatter across her breast like a pearl necklace
>
> Ye Qinghua (nineteenth century), "Tea Picking Women *zhuzhici*, #2"
>
> 纖纖十指採新茶　　頭戴涼棚面罩紗
> 火傘高張猶不息　　淋漓香汗似珠花
>
> 葉青華 《採茶女竹枝詞其二》[11]

Ye's poem takes special note of the challenging meteorological conditions of Taiwan—its intense tropical sun and sultry temperatures—which distinguished it, at least marginally, from his home in Zhejiang to the north. Just as tea plants and associated technologies had been imported from across the straits in relatively recent times,[12] these women descended from the waves of immigration from the nearby provinces of Fujian and Guangdong that peaked in the Qianlong and Jiaqing eras (1736–1820). The famed Taiwanese Hakka official and poet, Qiu Fengjia (1864–1912), who gained a reputation as a prodigy thanks to a *zhuzhici* he wrote for a prefectural examination at the age of twelve, employed botanical metaphors for the history of Han settlement.

> Chinese flowed in and settled on the land, their homes hardly more than nests
> Chaozhou, Huizhou, Zhangzhou, and Quanzhou people were most numerous
> After 200 years of spreading their tubers
> These small grasses that first alighted here have now set down roots
>
> Qiu Fengjia, "Taiwan zhuzhici"
>
> 唐山流寓話巢痕　　潮惠漳泉齒最繁
> 二百年來蕃衍後　　寄生小草已生根
>
> 丘逢甲,《台灣竹枝詞》[13]

Another poem by an obscure official of the same era serving in Taiwan alludes to the population of this ethnically heterogeneous society in similar terms:

> Clearing brush and uprooting weeds, together they planted hemp
> Barbarian fires surrounded their peaceful mock orange blossoms
> With nothing amiss in the landscape, their folkways plain and simple
> Their sons and grandsons hitched to the House of Han
>
> Wang Chaoqing (fl. 1880), "Zhuluo zhuzhici"
>
> 犵草除根共種麻　　蠻烟簇擁太平花
> 山川無恙民情樸　　子子孫孫屬漢家
>
> 王朝清,《諸羅竹枝詞》[14]

These poems and many others like them record the physical and social transformation of the landscape at the hands of Han migrants, casting this as the migration of both human and horticultural "settlers." As in other frontier regions, the influx of outsiders pushed out the Indigenous peoples' slash-and-burn agriculture and hunting, replacing them with wet-rice cultivation, sericulture, and various other labor-intensive, sedentary crops, including tea. Although not a few Taiwan *zhuzhici* portray the encounters between indigenes and migrants in more ambivalent terms, by and large the displacement of the former by the latter is praised for enhancing agricultural fecundity, and also for implanting a civilizing social order that suppressed the dreaded practice of headhunting (the decline of which was attributed to the folk hero Wu Feng, ca. 1730).[15] The maturation of the local pursuit of literature, and the Confucian classical training necessary to produce successful examination graduates and officials, eventually including metropolitan graduates (*jinshi*), were similarly celebrated as the happy fruit of the island's affixation to both the culture and the polity of the mainland. And it is as a visible symbol of this acculturation, through its ties to Chinese literati elite interests and concerns, that tea cultivation by tea-picking women emerges as a significant element of early modern Taiwan cultural formation, creating affective resonance between the cultural and natural elements of its landscape.

CULTURED TEA

April showers having just passed, we search for Maoluo
Tender leaves unfurl, mirroring emerald vines
My only regret is not having witnessed tea adepts compete
Spring breezes blew in vain for the song for picking tea

Wu Tinghua (1682–1755), "Aboriginal village *zhuzhici*"

纔過穀雨覓貓螺　　嫩葉旗槍映翠蘿
獨惜未經嫻茗戰　　春風辜負採茶歌

吳廷華《社寮竹枝詞》[16]

Wu Tinghua, an official interpreter (*tongshi*) between indigenes and Han during the early Qianlong reign, adds in an appended note to this poem that, although they harvested tea to trade for other items, the natives did not themselves drink it, because its nature was "cold."[17] The consumption of tea, it seems, was strictly a Han affair, identified with mainland migrants rather than the native aborigines, even if the latter were entrusted with harvesting it in the wild. The gradual transition from aboriginal to ethnic Han dominance over the island is also marked by the rise of tea as a crop that is not only cultivated but also consumed locally.[18] The "secure hitching" to the House of Han exuberantly noted in Wang Chaoqing's poem above may have been preceded by some reliance on the labor of the Indigenous population, but bringing tea under cultivation necessitated the gradual "taming" and eventual displacement of the indigenes, who were forced to retreat ever farther behind the so-called *jie* or pale that was—at least originally—intended to protect them from Han encroachment. Moreover, even if they did not directly uproot the natives, the spread of tea plantations into the mountainous interior of the island reduced the land available for hunting and other traditional pursuits. As in other regions of the world, over time these pressures had the effect of severely decimating the Indigenous population to a tiny fraction of their original numbers.

The transformation of the hills that resulted from the increase in tea cultivation is vividly portrayed in the following poem about the tea-covered hillsides near what is now the Taipei Zoo:

Wenshan, enveloped in emerald green, does not toot its own horn
Its dense tea plants' loveliness grows of itself
Unlike a weedy knoll whose flowers may indeed be radiant
But is suited to viewing for but a brief span of the entire year

Liu Yuying (ca. 1880), "Taibei zhuzhici"

文山挹翠不矜奇　　茶樹深深美自滋
不似草山花自豔　　四時只見一時宜

劉育英,《臺北竹枝詞》[19]

This poem sets Wenshan, literally a "cultivated, refined mount" in opposition to Caoshan, a "weedy knoll." The latter is covered by wildflowers that are seductively brilliant but undisciplined and ephemeral, the former by tea that though restrained and modest, pleases its visitors the whole year-round. This juxtaposition implies that the cultivation of tea introduces a level of civilization, or civility (*wen*), as well as a permanence that trumps the naturally appealing but slightly unkempt and less durable native beauty of the landscape.

That tea production eventually became wedded to Han ethnicity is also evident in the following poem about tea-picking maidens:

Going down the mountain to pick tea, the sun is about to set
Going up the mountain to cut wood, the clouds are not yet dense
Whose daughter could be singing that folksong?
Her song sounds half Zhangzhou dialect, and half Cantonese

下山採茶日欲沉　　上山伐木雲未深
山歌一曲誰家女　　半帶漳音半粵音。

莊嵩 (ca. 1920)《台中竹枝詞》[20]

For Ye Qinghua, ethnic identity is not quite as overtly expressed as this, but the demure coyness of tea-picking maidens marks them subtly but unmistakably as Han, as opposed to the seemingly (at least to Han eyes) less chaste indigene:

A good girl, she doesn't flirt
Her face peachy pink, her waist willowy
Light songs and words catch her listener's hearts
Even though they cannot see her cheeks' rising blush

好個娘兒不撒嬌　　桃花如面柳如腰
輕歌作語撩人意　　未見腮邊泛赤潮

When her song stops, she turns to calling out
Her baskets full, she walks hand in hand with others
We know they prance to show their happiness
And cast aside the boredom of their lonely time on the mountain

Ye Qinghua, "Tea-picker's zhuzhici #3, 4"

歌聲歇處轉傳呼　　滿篋盈筐接手扶
知是開心行載舞　　將丟悶緑一山孤

葉青華 《採茶女竹枝詞其三、四》[21]

The full flowering of tea cultivation, like the more abstract cultivation of poetry and other literati arts, had proceeded hand in hand, even if Han men occasionally strayed into the arms of the local Indigenous natives. This golden age of peace was relatively short-lived, though. Before long, precious metals and industrial combustibles grew to dominate the minds of many laborers and literati alike, including some of the poets of Taiwan *zhuzhici*. And their poems allude to both the strife and the pollution that such changes brought to the verdant tea-cloaked hills of the island.

BEYOND TEA

In addition to tea, many *zhuzhici* describe agricultural, forestry, marine, or other products of the island, praising their beauty, quality, uniqueness, or abundance. These include *shijia* (annona squamosa), sugar cane, and mango, or *wuyuzi* (*karasumi* or salted mullet roe), crocodile, and camphor wood. A small number are also devoted to another category of local production: the mining of gold and coal, especially in the north, that played a pivotal role in Taiwan's political and economic history after 1870. In a well-known *jueju* about gold mines by Qiu Fengjia, the mineral deposits of the Jilong area are portrayed as, at best, a mixed blessing:

The shadow of clouds extends over the side of Chicken Coop Mountain
Painstakingly sifting through the sands, [prospectors] wade in deep water
This treasure cache still remains, but its owner has changed thrice
Truly, there is unpropitious gold in the world of man

Qiu Fengjia, "Jilong Gold"

雞籠山畔陳雲陰　　辛苦披沙一水深

寶藏尚存三易主　　人間真有不祥金

丘逢甲《雞籠金》[22]

The title to the gold mines of Jilong—a synecdoche for Taiwan as a whole—has been transferred from the Indigenous people, to the Qing, and now (ca. 1900) into Japanese hands. Their gold arouses the greed of all who covet this land, and causes misfortune—an allusion to the unluckiness of gold in the Daoist text *Laozi*—as it draws both native adventurers and foreign intruders to the island. Other poems of this period, as well, describe the frenetic activities and desires unleashed by the exploitation of gold and coal deposits, the recovery of which requires that mountains be defaced, and streambeds be diverted, polluted, or filled, all in the rush to grab a share of the Earth's bounty before it is claimed by others.

Gold and copper ore are both superlative
The vein of their ore is said to reach all the way to Fuzhou
Flowering trees here grow very sparsely, and vegetables are few
Over the whole mountain, only the grass looks soft and inviting

金銅礦石兩稱優　　地脈相傳透福州
花木極稀蔬菜少　　遍山唯見草纖柔

The hopes of all are pinned on becoming an instant millionaire
Prospectors for gold can't bear poverty
Already, their ranks have swelled to over 30,000 mouths
They pray to the local earth gods to grant them luck for riches

Cai Qingrong (ca. 1900), "Ruifang mines zhuzhici"

均望須於作富翁　　菜金人不認貧窮
已超三萬餘人口　　福運爭求土地公

蔡青榕,《瑞芳鑛山竹枝詞》[23]

Such gold- and coal-themed poems can be seen as occupying one end of a spectrum of poems about local products (*wuchan*) for which the cultivation of tea lies at another end. Many of the mining poems recount the dogged efforts of the miners, their hardships, the overcoming of dangers or other obstacles, and the sheer thrill of striking it rich. In these poems, mining and its protagonists are presented in a heroic, swashbuckling, decidedly masculine mode, sometimes tinged with ambivalence, that makes for a striking contrast with the alluringly sensuous descriptions of tea-picking maidens amid verdant hillsides of the tea-themed poems. In tea-picking *zhuzhici* such as that by Ye Qinghua cited above, it is not the lure of wealth, but the loveliness of the landscape, and the care, harvesting, and production of the plant, that is memorialized. The figure of the lithe, lissome young tea picker (identified both as *caichanü*, or simply *niang*) thus emerges as a foil, mostly implicitly, to the foolhardy miners who risk life and limb in search of treasure.

Some tea poems are written about the very same areas as those of the mines, to the north and east of Taipei. And in at least one of these, set in Xizhi (20 km east of Taipei), these two ways of life come into direct conflict:

Rice fields and tea hills fill one's view
This terrain is naturally suited to farming
But once coalmines started to open up one after another
Countless tea plantations have been fouled and abandoned

Gao Huiran, "Tanyin zhuzhici"

田稻山茶一堵望　　天然地勢好農場
自從炭礦頻開發　　無數茶園任穢荒

高惠然 (ca. 1890) 《灘音竹枝詞》[24]

Descriptions of the competition or contradictions between these two economic spheres are relatively uncommon in *zhuzhici*, but the vulnerability of tea cultivation to economic changes, and correspondingly of the women who work in the fields, is evident in other poems, such as this one about the decline of Xinzhu tea plantations at the end of the era of Japanese rule:

Tea plantations everywhere are overgrown and abandoned
Tender young girls left without work can only lament
As the sun sets, we try to listen along the road behind the cars
How many households are still singing tea-picking songs?

Xie Jingyun (ca. 1950), "Xinzhu zhuzhici"

茶園處處荒廢多　　失業嬌娃恨若何
日暮試聽後車路　　幾家人唱採茶歌

謝景雲,《新竹竹枝詞》[25]

Due to a combination of wartime disruption, restrictions by the colonial authorities, and competition among tea producers in South and East Asia, Taiwan tea production suffered declines through the mid-1940s, which may explain this lament about unemployment. And not only for tea, but in agriculture, fisheries, and other industries, the precarity and dangers faced by the laboring classes was evident in early twentieth-century *zhuzhici* on multiple topics. Concern for the difficult conditions and unpredictable fates of tea laborers is even discernible in Japanese haiku and tanka poems written during Japanese rule (1895–1945).[26]

In fact, we might go as far as to say that these *zhuzhici* chronicling the difficulties of the Japanese era give voice to what the Australian philosopher Glen Albrecht has termed *solastalgia*, a kind of psychosocial reaction to societal or environmental change. It is triggered by "the loss of a secure environment . . . a form of homesickness one experiences when one is still at 'home' . . . a feeling of dislocation; of being undermined by forces that destroy the potential for solace to be derived from the immediate and given."[27] Although this concept has been applied most often to relatively recent conditions associated with climate change, Albrecht originally coined it to describe the psychosocial and somatic effects observed in local residents who lived near open-pit coalmines in a region of south-central Queensland, Australia during the 1980s–1990s. And thus, it seems potentially relevant to examining the literary responses to environmental and social changes taking place in East Asia, including Taiwan, in the late-nineteenth century and beyond.

We can push this solastalgic impulse back even further into the past, prior to the displacement of tea workers by mining or industry as lamented by Gao Huiran and Xie Jingyun. As already noted, *zhuzhici* were employed not infrequently to chronicle changes wrought through settlement, development, or expansion, both in Taiwan and elsewhere. Traditionally conceived as the adaptations of folksongs by literati authors through the ancient practice of *caifeng* ("collecting airs"), they represent observations of the ever-fluid state of customs and mores of society, and the natural environments that may be affected by or in turn affect them. Tea-themed as well as other *zhuzhici* provide a unique medium for commentary about such transformations, since they typically adopt the voice of native informants, and especially of the lowly, uneducated, and female among them.

Solastalgia may not seem quite the right term to apply to the early modern era before the more momentous disruptions of the late-twentieth and early twenty-first centuries. Nevertheless, the sense of disorientation caused by rapid development like that of late-nineteenth-century Taipei seems akin if not entirely identical to the loss of familiarity with and thus of solace obtained from one's home environment as described by Albrecht. Even material prosperity, especially if it alters the landscape, can engender feelings of perplexity or loss. And while such emotional responses to unsettling changes are a staple of the poetic tradition going back to the *Shijing* (e.g., "*Shuli*," #102), *zhuzhici* seem to have been especially well-suited to such themes, particularly as the impact of modernity grew ever more disorienting and alienating. Hence, we of today might learn from these poets of the relatively recent past, and give voice to the sense of foreboding and loss that the recent breaching of 420 ppm level of atmospheric CO_2 promises to impose on us, our children, and future generations.

NOTES

1. All the poems cited in this chapter were published before 1982, thus belonging to the public domain and permissible to be cited in full length. I thank the editor of Lexington for their helpful reminder.

2. For an overview of arguments advocating for the insertion of the embodied subject into academic writing through "tack[ing] between the analytical and the sensible," between "embodied form as well as disembodied logic," see Paul Stoller, *Sensuous Scholarship* (Philadelphia: University of Pennsylvania Press, 1997), xiii and 2–24.

3. Silvia Gherardi, "Theorizing Affective Ethnography for Organization Studies," *Organization* 26, no. 6 (2019): 745–754.

4. Lone Bertelsen and Andrew Murphie, "An Ethics of Everyday Infinities and Powers: Félix Guattari on Affect and the Refrain," in *The Affect Theory Reader*, edited by Melissa Gregg and Gregory J. Seigworth (Durham, NC: Duke University Press, 2010), 146.

5. Gasi Islam and Michael J. Zyphur, "The Sweetest Dreams That Labor Knows: Robert Al and the Poetics of Work," *Management Decision* 44, no. 4 (2006): 527.

6. Lone Bertelsen and Andrew Murphie, "An Ethics of Everyday Infinities and Powers: Félix Guattari on Affect and the Refrain," 151.

7. Qi Gao, *Gao Qingqiu ji* (Shanghai: Shanghai guji chubanshe, 1985), vol. 1, 84.

8. Liqi Wang, Shenzhi Wang, and Zijin Wang, eds., *Lidai zhuzhici* (Xi'an: Shaanxi renmin chubanshe, 2003), vol. 1, 520.

9. Shengfeng Weng, *Qingdai Taiwan zhuzhici zhi yanjiu* (Taipei: Wenjin chubanshe, 1996), 280ff.

10. Jie Sun, *Zhuzhici fazhanshi* (Shanghai: Shiji chubanshe, 2014), 27–30, 132, 198.

11. Xiang Chen, *Taiwan zhuzhici xuanji* (Taipei: Taiwan Shangwu yinshuguan, 2006), 241.

12. See Naoto Kawarabayashi, *Kindai Ajia to Taiwan: Taiwan chagyō no rekishiteki tenkai* (Kyoto: Sekai shisōsha, 2003), 18–50.

13. Xianzong Gong, ed., *Taiwan zhuzhici sanbaishou* (Tainan: Kailang zazhi, 2010), 34.

14. Chen, *zhuzhici*, 123.

15. Mei'e Huang, "Kuaguo Moshuigou de Han wenxue: Ming Qing shiqi Taiwan gudian wenxue," 45; also Kanghu Jiang, "Wu Feng Miao Mao," in *Taiwan yuanzhuminzu guanxi wenxue zuopin xuanji* (1895–1945), ed. Mei'e Huang (Tainan: Taiwan wenxueguan, 2008), 87.

16. Chen, *zhuzhici*, 135.

17. Chen, *zhuzhici*, 135.

18. For a discussion of Qing views of Taiwan's gradual transformation from Indigenous—to Han—majority, see Emma Jinhua Teng, *Taiwan's Imagined Geography* (Cambridge, MA: Harvard University Press, 2004), esp. chapters 1, 4.

19. Chen, *zhuzhici*, 158–9.

20. Chen, *zhuzhici*, 114.

21. Chen, *zhuzhici*, 252.

22. Qingsong Chen, ed., *Jilong gudian wenxueshi* (Keelung, Taiwan: Jilong wenhuaju, 2004), 142.

23. Chen, *zhuzhici*, 175–176.

24. Chen, *zhuzhici*, 172.

25. Chen, *zhuzhici*, 134.

26. Mei'e Huang, ed., *Taiwan yuanzhuminzu guanxi wenxue zuopin mulu 1603–1945*, 29–45.

27. Glenn Albrecht, "Solastalgia: The Distress Caused by Environmental Change," *Australasian Psychiatry* 15, no.1 (2007): 95–98.

BIBLIOGRAPHY

Albrecht, Glenn. "Solastalgia: the Distress Caused by Environmental Change." *Australasian Psychiatry* 15, no.1 (2007): 95–98.

Bertelsen, Lone and Andrew Murphie. "An Ethics of Everyday Infinities and Powers: Félix Guattari on Affect and the Refrain." In *The Affect Theory Reader*, edited by Melissa Gregg and Gregory J. Seigworth, 146–160. Durham, NC: Duke University Press, 2010.

Chen, Qingsong 陳青松, ed. *Jilong gudian wenxueshi* 基隆古典文學史. [A History of Classical Literature of Jilong]. Keelung, Taiwan: Jilong wenhuaju, 2004.

Chen, Xiang (陳香). *Taiwan zhuzhici xuanji* 台灣竹枝詞選集. Taipei: Taiwan Shangwu yinshuguan, 2006.

Gao, Qi (高啟). *Gao Qingqiu ji* 高青丘集 [Collected Works of Gao Qiu]. Shanghai: Shanghai guji chubanshe, 1985.

Gherardi, Silvia. "Theorizing Affective Ethnography for Organization Studies." *Organization* 26, no. 6 (2019): 741–760.

Gong, Xianzong (龔顯宗), ed. *Taiwan zhuzhici sanbaishou* 台灣竹枝詞三百首[300 Taiwan Bamboo Branch Lyrics]. Tainan: Kailang zazhi, 2010.

Huang, Mei'e (黃美娥). "Kuaguo Moshuigou de Han wenxue: Ming Qing shiqi Taiwan gudian wenxue." 跨國墨水溝的漢文學：明清時期台灣古典文學 [Chinese Literature Spanning the Black Current National Divide: Classical Taiwanese Literature of the Ming and Qing Eras]. In *Wenxue@Taiwan: 11 wei xinrui Taiwan wenxue yanjiuzhe* 文學@台灣： 11位新銳台灣文學研究者 [Literature @ Taiwan: 11 Promising Young Scholars of Taiwan Literature]. Tainan: Taiwan wenxueguan, 2008.

Islam, Gasi and Michael J. Zyphur. "The Sweetest Dreams that Labor Knows: Robert Frost and the Poetics of Work." *Management Decision* 44, no. 4 (2006): 526–535.

Jiang, Kanghu (江亢虎). "Wu Feng Miao Mao." 吳鳳廟貌 [The Condition of the Temple to Wu Feng] In *Taiwan yuanzhuminzu guanxi wenxue zuopin xuanji* 台灣原住民族關係文學作品選集 (1895–1945) [Selected Literary Works Related to Taiwan's Aborigines], edited by Huang Mei'e 黃美娥, 87. Taipei: Xingzhengyuan yuanzhumin weiyuanhui, 2013.

Kawarabayashi, Naoto (河原林直人). *Kindai Ajia to Taiwan: Taiwan chagyō no rekishiteki tenkai* 近代アジアと台湾：台湾茶業の歴史的展開 [Taiwan and Modern Asia: The Historical Development of the Taiwan Tea Industry]. Kyoto: Sekai shisōsha, 2003.

Qiu, Liangren (丘良任), Chao Pan (潘超), Zhongquan Sun (孫忠銓), and Jin Qiu (丘進). *Zhonghua zhuzhici quanbian* (中華竹枝詞全編) [Complete Bamboo Branch Lyrics of China]. Beijing: Beijing chubanshe, 2007.

Stoller, Paul. *Sensuous Scholarship*. Philadelphia: University of Pennsylvania Press, 1997.

Sun, Jie (孫杰). *Zhuzhici fazhanshi* 竹枝詞發展史 [History of the Development of Bamboo Branch Lyrics]. Shanghai: Shiji chubanshe, 2014.

Wang, Liqi (王利器), Shenzhi Wang (王慎之) and Zijin Wang (王子今). *Lidai zhuzhici* 歷代竹枝詞 [Bamboo Branch Lyrics of Multiple Eras]. 5 Vols. Xi'an: Shaanxi renmin chubanshe, 2003.

Weng, Shengfeng (翁聖峰). *Qingdai Taiwan zhuzhici zhi yanjiu* 清代台灣竹枝詞之研究 [Qing Dynasty Bamboo Branch Lyrics of Taiwan]. Taipei: Wenjin chubanshe, 1996.

Chapter 5

Brothers Locked in Strife

A Molecular Reading of Blood Camphor

Ya-feng Wu

Camphor products, together with tea and sugar, had been the three major export commodities of Taiwan, from the final decades of the Qing dynasty through Japanese colonial regime to the early decades of Nationalist government rule on the island. Its peak production was based upon deforestation of the mid-level mountains and dislocation of the Indigenous tribes, mainly the Atayal. The bloody confrontations between the Han Chinese settlers and the Indigenous peoples, later compounded with the Japanese police force, were the backdrop to what the government propaganda claims as the miracle of the "Kingdom of Camphor."[1] Camphor provides a figure through which we can better gauge the condition of Taiwan caught in imperial contentions. Perhaps because the experience is still raw, it has seldom been treated as a major literary theme. *Blood Camphor* (2017) by Ying-Min Chang (張英珉) is the first significant exception. The novel is set in the most unsettling time (1855-c.1875), straddling between the troubles of the 1868 Camphor War, in which the British secured privileges from the Qing court to procure camphor products from Taiwan, and ending about two decades before Japanese colonial rule in 1895. Chang effectively chooses to focus on an era when intense international scramble for natural resources of Taiwan met no clear governing regime. In other words, Chang's novel refers to a period in which the forest of Taiwan was still a site of multiple-colonization, literally an extreme case of sub-colony, before it became subjected to the development model of "state hegemony," a dire condition which has been investigated by Kao Jun-Honn.[2] Furthermore, Chang designs a framing device that bookends the novel with accounts of the conflict taken from the perspective of the ants, featuring the two leaders of the Han Chinese and the Atayal lying paralyzed beside camphor trees. These two turn out to be brothers brought up by the rival communities respectively. Their maimed bodies with ants crawling on them not only form an emblem of the conflict but also present a non-human perspective on human wrangling. This mechanism evokes a Deleuzean molecular interpretation of human affairs that "guarantee[s]" a "continuum" through "dissolution" of definite forms.[3] In addition, the titular focus on camphor products requires an approach drawn from Arjun Appadurai's *The Social Life of Things* to assess the meanings of objects as "inscribed in their forms, their uses, [and] their trajectories" and to believe that "it is the things-in-motion that illuminate their human and social context."[4] This approach helps us to see how the history of Taiwan and camphor trees had been intertwined as both gained an increasingly important strategic significance in the nineteenth and twentieth centuries.

Chang's novel begins in the year 1855 when one big military conflict took place between the Han Chinese settlers and the Atayal clans. Two Chinese infants were left in the faggot baskets for their parents believed the rumor that the Atayal warriors though ferocious would not harm infants. The Atayal leader, Buni, took the elder infant home, who grew up to be Walis. The other infant, Wen-shi Li, was later adopted by Hsiu-he Liu, the master of Chinese settlement. The novel unfolds as the two brothers developed to become chief of their community. Under the increasing demands from imperial merchants for camphor products, the Han settlers and the Atayal clans fought with each other, for the former to maximize camphor manufacturing and for the latter to preserve ancestral hunting grounds. This chapter examines how Chang's novel dramatizes the inextricable relation between Taiwan and camphor, by placing camphor products under the framework of Appadurai's social history of commodity and by revealing a Deleuzean molecular structure that is intimated in the framing device, in order to enable the reader to participate in a molecular experience with the humans and nonhumans caught up in the strife.

QING CHINA'S IMAGINED GEOGRAPHY

The situation of Taiwan needs to be understood in a concurrent connection with China. Emma Jinhua Teng traces how Taiwan became part of the "imagined geography" of Qing China during the period 1683–1895 in examining Taiwan's transformation from a "savage island" located "beyond the seas" to an integral part of the Chinese multiethnic empire.[5] Chang's novel deals with one of the most volatile periods in this process of fashioning the "imagined geography" when Qing began to realize the vital role that Taiwan would play in international relations.

Taiwan became a prefecture under Fujien in 1684. As Teng points out, in the Ming dynasty, Taiwan was characterized as "the ball of mud," whereas in the eighteenth century Taiwan was "denigrated and idealized" as a "living museum of Chinese antiquity," and in the nineteenth century it became a "land of Green Gold" (namely timber).[6] Taiwan's status had been affected by Qing's dealing with the Indigenous peoples and natural resources of the island. John Shepherd describes Qing's policy of Taiwan as vacillating between a pro-quarantine and a pro-colonizing approach.[7] It is all the more dramatically exemplified in its policy toward the Indigenous peoples. Qing court had long adopted a practice dating back to the Song dynasty concerning non-Han peoples, which was to divide them into "the cooked" and "the raw."[8] The cooked Indigenous peoples were used as a "buffer" between the raw Indigenous tribes and the Han Chinese settlers.[9] The defining criterion between these two groups was always fluctuating depending on the strategic necessity of the Qing. This roughly speaking nonchalant policy was gravely challenged by two events: the Rover Incident (1867) and the Mudanshe Incident (1871), which forced the Qing court to realize that Taiwan had become one hotspot due to its geopolitical position and abundant resources. They could no longer shrug off responsibility and wave away invaders by claiming the Indigenous peoples of the island lying outside their jurisdiction. More specifically, the Qing court implemented the "Open the Mountains and Pacify the Savages" policy in 1875 with a view to legalizing the advancement of Han Chinese settlers into the Indigenous territory.[10] It finally declared Taiwan a province of the Qing China in 1885. During this long process of "geographical imagination," different parts of Taiwan were assessed differently by the Qing court: the plains were most valued when the island was claimed to be the "Granary of China" (for rice and sugar); whereas the upland forested areas changed from a source of disease and threat to become valued for tea and camphor products.[11] In other words, geographical imagination joined economic exploitation to form the basis of empire building.

INTERTWINED HISTORY: CAMPHOR TRADE AND TAIWAN

Earliest record of trade in camphor in the East Asian area, as Ronald C. Po points out, can be traced to the eighth century.[12] In Chinese *materia medica*, camphor products had been recommended for respiratory complaints, rheumatism, toothache, disinfection, and so on.[13] This set of knowledge was transmitted via Arabian merchants to Europe and the demand increased at the onset of the Black Death for they were used as fumigants against the polluted air.[14] Prior to the 1870s, camphor products were mainly high-end luxury goods meant for medicinal and aromatic use.[15] The rise of the celluloid industry since the 1870s, which used camphor products as solvent and plasticizer, changed them to raw materials in need of mass production.[16] This phenomenon precipitated the creation of artificial camphor products and later the full synthetization was developed in 1896.[17] But as Rebecca Altman points out, celluloid industry increased, rather than decreased, the demand for camphor.[18] In effect, Taiwan remained the major exporter of camphor products in the world till the 1920s.[19]

Camphor stills,[20] which used to be commonly found in the mountainous areas of northern Taiwan, were gradually replaced by tea plantations and then by recreational farms and theme parks. We have forgotten this bloody chapter of history that thrusts Taiwan into the global capitalist network. Camphor has been regarded as an apt figure for Taiwan as in *A Man Who Has Been through Three Ages* (2010) by Shu-ching Shih (施叔青), in which a camphor tree speaks on its own accord as a witness to the fate of the inhabitants in Taiwan. Shih's focus is placed mainly on Han settlers. Chang's novel, however, provides a contrapuntal supplement to Shih's by pairing up the Atayal clans with the Han Chinese settlers in their struggle to survive imperial extraction. Moreover, Chang's two-fold story is augmented by a framing device with a perspective from the ants as a foil to human affairs.

The fates of Taiwan and camphor are indeed inseparable. As R. A. Donkin points out, the camphor trees (*Cinnamomum camphora*) are indigenous to eastern China (from Kiangsu to Kwangtung and Hai-nan, extending west to Szechwan and Yunnan) and its "geo-botanical center of diversity" lies in Chekiang, Fukien, and Taiwan.[21] Nowadays, botanists have changed their academic nomenclature to *Camphora officinarum*, making them distinct from cinnamon.[22] The camphor trees of Taiwan were first remarked by the Portuguese Jesuit Alvarez Semedo (c. 1633) and the earliest reference to the production on the island is found in H. J. von Klaproth's *Mémoires* (1826).[23] As camphor trees were gradually depleted on the coast of Fukien, Taiwan became an increasingly important center of production. In 1903, James W. Davidson, the first American Consul to the island, revealed his deeply-felt insight on the ramifications of this global trade in *The Island of Formosa, Past and Present*:

> [t]he fact that [camphor] is snatched from the jungle over which the wild savage roams, and that it is not produced to any extent in any other part of the world, save Japan, accounts for this. It would be an inviting subject for the statistician . . . to figure out how many drops of human blood are represented in the few ounces of Camphor which the humane young lady purchases to keep her dainty garments free of moths, or how many lives are lost that some decrepid [*sic*] old gentleman may be cured of his rheumatic pains.[24]

Davidson captures the moral dilemma faced by someone like himself who could afford to observe the sorry devastation caused by insatiable desires of the Western market. As Chih-Chieh Yu demonstrates, the former abundant virgin camphor forests of Taiwan, once marveled by William Robert Price in 1912, have largely disappeared under the state development scheme, including Japanese colonial regime and the Nationalist government after the Second World War.[25]

Moreover, the evolution and study of the camphor industry serve as an index to the overall development of Taiwan. Antonio C. Tavares provides an overview of the scholarship on the camphor industry. Earliest studies done during the Japanese colonial period had presented "a progressive narrative of the expansion of civilization" and "state power over an industry and frontier," whereas early post-war and nationalist government studies emphasized the conflicts between Qing and British agents.[26] Later, a revisionist study of the 1970s highlighted a "neo-classical economic analysis of the growth of the camphor trade."[27] The 1990s saw a new interest in Chinese-Indigenous relations promoted by emerging identity politics.[28] Tavares's studies builds on a keen awareness of the ideological strains of previous scholarship to emphasize the way in which in the 1860–1895 period "Qing increased its reconfiguration of institutions, commercial networks, and practices governing the camphor trade as Taiwan was connected to the world market via a contradictory process of imperial engagements between Qing and British empires on the China coast."[29] Tavares's stance and argument help us to see how Chang enlivens the complicated historical background to build a foundation for empathy, rather than for recrimination.

Camphor trees play a significant part in the Atayal life although scholars have yet to fully explore the cultural and mythical implications of camphor in the Indigenous peoples' history. A contemporary Atayal writer, Walis Nokan (瓦歷斯・ 諾幹), offers us an insight in his essay "Telephone Receiver." As Walis Nokan describes, in the past the Atayal traders would tap the camphor tree trunk to send messages to their fellow clansmen concerning the business deal they struck with the Hakka settlement. If the deal was successful, the trembling tree trunk told their clansmen the toad on their skin would be repelled by salt bought from the Hakka. If the Hakka settlers cheated them, the tapping sound from the tree trunk would alert the Atayal to get ready for revenge. It was done as if the camphor tree sent out words to the waiting clansmen, and as if their footprints were inscribed on the skins of the forest.[30] Walis Nokan attempts to describe the way in which the Indigenous peoples developed their linguistic communication including textile patterns, oral language, etc. through their ingenious cooperation with camphor trees. This passage in fact reminds us not only of the volatile and even treacherous nature of trade between various groups but more importantly how the habitat of camphor trees, which facilitates cooperation between humans and nonhumans, has been destroyed by Han Chinese's extracting of camphor resources.

CAMPHOR WAR AND THE NOVEL

Camphor War stands for a landmark in the long process of imperial schemes concerning the camphor products and the region. Ronald C. Po focuses on the events leading up to the Camphor War and its ramifications. He especially mentions the significant role played by three foreign "agents": Dr. William Jardine of the Jardine Matheson Co., William Alexander Pickering the British officer, and Charles W. Le Gendre of the American Consul at Amoy. Defeated in the war, the Qing court was forced to relinquish its monopoly, but the British minister in Beijing, Sir Rutherford Alcock, was later forced to return the indemnity, which had been imposed upon the Qing court.[31] Po's article helps us to see how Chang molds the British merchant, Gibson, in the novel, against the prototype of these foreign officials. Furthermore, the Camphor War was staged under the banner of free trade on the British side. Li Pei-chen reminds us that in the period 1850–1868 before the Camphor War both Chinese and British merchants shifted their abiding principle between monopoly and free trade according to their business interests at the time. In other words, all parties competed to buy camphor products from Han Chinese settlers at higher prices,[32] regardless of their respective ideological persuasion. Such an unstable situation is vividly captured in the novel. Gibson told Master Liu via an interpreter (who was from Canton and could speak Hakka in addition to English) that the British were buying camphor products from Taiwan and then selling them to the Americans who would make them into medicine of a much higher value.[33] Gibson promised to pay three times higher and demanded a large quantity. In order to meet this steep demand, Liu's adopted son Li went downstream to recruit 100 more workers with the promise of higher wages in order to offset the fear of the headhunting Indigenous peoples and unknown diseases. The intensified global demand for camphor products inevitably accelerated the conflict between the Han Chinese and the Atayal.

The novel thus thematizes the inextricable history of Taiwan and camphor. Chang incorporates what Taveras calls "a motley crew of players"[34] who were active in the region before and in the wake of the Camphor War, including the Han Chinese settlers, the Indigenous peoples, the foreign merchants, and the mixed-blood interpreters. Chang masters the pulse of historical changes in making his plot evolve with the swiftly developing camphor industry. Chang anchors the story on the historical conflict between the Han Chinese settlement Ching Tong Fu (金東福) and the Atayal clans, Wang Tz Fan (王字番),

along the line of camphor checkpoints (隘勇線).[35] The name Ching Tong Fu suggests an echo of the historical settlement Ching Kuang Fu (金廣福), an alliance between Hakka and Hoklo groups which was sanctioned by the Qing administration.[36] The elements of history and fiction mirror and refract each other to produce new insights out of the fractures and fissures between fact and literary effect.

In the fashion of a chronicle, the novel begins with the specific date of 1855,[37] when Buni and his cousin Bulan invaded New Village. They discovered two infants in the baskets of faggots left by the Han Chinese, the older one reaching out to Buni. He took him home to his wife who just lost her baby in a difficult birth. The novel unfolds along the growing up of these two babies until they reached about sixteen years of age. The Chinese settlers gradually set up camphor stills beyond the boundary that was sanctioned by Qing's imperial policy. Master Liu said if they had the choice they would have stayed within the pails and contented with farming and silkworm raising, which were the traditional Chinese country life.[38] Instead, they were tempted by the high prices of camphor products offered by international merchants. No wonder the British merchant smirked in thinking that they would prefer trade to colonization to maximize their profit without incurring unwanted responsibility and risk.[39] The Atayal clan, thus bitterly sandwiched between the Han settlement and other Atayal communities, was forced to defend their ancestral hunting ground.

THE TWO BROTHERS AND THE HEADHUNTING RITUAL

The upbringing of the two brothers amounts to an experiment in the scheme of nature versus culture, but situated in the great struggle for survival. The younger brother Li, who was brought up by the Chinese martial master, is well-versed in the *Art of War*,[40] especially psychological maneuver, whereas the elder brother Walis, who was brought up by the Atayal, inherited the stratagem and agility most suitable for mountainous terrains. On the one hand, Walis' tactics is to harass the sentinels without hurting the camphor workers so that their manufacturing would be delayed. It all bears down to business acumen. He hoped in the end the Han Chinese would realize it was too costly to develop the camphor industry.[41] On the other, Li's tactic was to negotiate with the Atayal by capturing a female Atayal and using her to trade for six-months peace so that they could speed up the production and he was even willing to share the profit with the Atayal.[42] The two brothers are indeed comparable in valor and resolution. They schemed against each other only to die in similarly pyrrhic waste.

In effect, the two brothers materialize the fate of the Han Chinese and the Atayal who were both pawns to imperial exploiters.[43] Some of the Han Chinese settlers came to Taiwan to find a better life away from the troubles in the wake of the Taiping Heavenly Kingdom Movement (1851–1864). Some followers of the Hakka founder of this Christian theocratic organization, Xiuquan Hong (洪秀全), such as Master Liu and his entourage,[44] came to Taiwan and found themselves pushed to the mountainous fringes by the Hoklo settlers who had come from Fukien earlier. Liu was plagued with grief and guilt for he had lost a few wives and children to miasma and lost his comrades in their fight with the Indigenous peoples. He turned to opium which was provided by the British merchant with a view to expediting the deal on camphor products.[45] Interestingly, Liu understood what Gibson and his ilk would enjoy by organizing a spectacle of Chinese workers chipping at camphor trunks and by showing native animal species for them to facilitate their impersonation of colonial officers like Robert Swinhoe.[46] The curating of such an imperialist spectacle suggests the novelist's self-conscious humane irony on both the Han Chinese settlers and the foreign merchants.

Some veterans fighting against Hong also came to Taiwan because they did not have the prospect of promotion due to their not having ties with the main official force, the Xiang Battalion (湘軍).[47] In Taiwan, they helped their employers to build New Village and guard the camphor checkpoint, which was the boundary set up by the Qing administration to separate the Han Chinese and the Indigenous peoples and also to exempt itself from responsibility if the Han Chinese ventured into uncharted territory. But it was the boundary that the Han Chinese settlers were impelled to push beyond in order to maximize their camphor extraction to meet the increasing demands of British, American, and German imperial clients. If they went beyond the boundary, they were at the mercy of the Indigenous peoples who bore grudges against the Han settlers for impinging on their ancestral hunting grounds. Meanwhile, the Atayal tribes were also in dispute with each other over old scores. In other words, the global trade of camphor products rendered this region a bloody battlefield.

The Atayal clan, under Walis's guidance, kept on harassing the Chinese workers with the hope that they would give up extracting camphor.[48] They did not consider this sort of attack headhunting. The only true headhunting in the novel was done by Buni and his son Walis with the view to proving Walis's Atayal identity and adulthood.[49] They went close to the Chinese quarter and witnessed how they chiseled the camphor trunk.[50] To them, the Han Chinese were no other than devils who reduced their trees to dust. They did not realize the gunpowder they acquired from the Han or the British would also contain camphor products.

An earlier "accidental" headhunting occurred when Nogen's father went over the river boundary to retrieve a costly metal trap but was beheaded by Walis's father and his cousin. Even if Walis's clan had observed the ritual to welcome Nogen's father's spirit to live with them,[51] the ritual could never appease Nogen. Eventually, it was Nogen who tricked Walis into believing that he had got Nogen's support in their invasion of New Village. In fact, Nogen cut a deal with Li through an interpreter, and therefore, Nogen was indirectly contributing to Walis's downfall.[52] Nogen himself was naïve enough to believe Li would yield Walis's hunting ground to his clan instead. Walis seemed to be alone in realizing their common destiny that "you are our frontline, we are your frontline, it is the fate of all Atayal."[53]

Headhunting as an Indigenous ritual has been examined thoroughly by anthropologists. According to Janet Hoskins, headhunting ritual was historically practiced "from as far north as mountainous parts of Taiwan, as far west as Assam, and southwards to the Indonesian archipelago."[54] It is an "organized, coherent form of violence in which the severed head is given a . . . ritual meaning and the act of headhunting is consecrated."[55] In addition, Scott Simon argues that headhunting amounts to "a ritual of community cohesiveness mediating contradictions between life and death, the familiar and the foreign, and human and non-human."[56] It was also part of the fertility ritual for only men who had captured heads could tattoo their faces, which made them attractive to potential wives.[57] They would drink from the skull with its blood, believing this would strengthen their prowess. In some communities, the head would be transformed into a friend and brother; some asked it to become an ancestor.[58] Nevertheless, the contradictions between headhunting as "the implementation of Gaya (sacred law) and consolidation of political power" eventually led to its demise.[59] These anthropological studies indeed pave the way for a different understanding of such crude but consecrated violence. Nevertheless, Chang honestly questions the validity of headhunting rituals because it simply bears down to a vicious cycle of revenge. Violence would never be assuaged. For both the Atayal and the Chinese, not being able to be buried intact is the greatest shame. In the novel, the cruelty of beheading is unflinchingly matched with the Chinese custom of selling and cooking Indigenous peoples' flesh. Some rumor had it that eating the Indigenous peoples' flesh would protect them from being headhunted.[60] The mixed-blood interpreters by chance went past the scene of flesh cooking and at once fled in terror.[61] The flesh merchant even exported his product to China,[62] which aggravated the macabre absurdity of human flesh trade under the guise of atavistic superstition.

THE MOLECULAR MODEL: THE ANTS' REFRAIN

Violence of such an intensity demands an understanding that does not seek to connive at its inhumanity by simply seeking to transcend tribal affiliation and individual subjectivity. Gilles Deleuze and Felix Guattari's idea of the molecular may help in formulating a perspective that is at once immanent and interpenetrative which helps us to contemplate redemption. The molecular, as Deleuze and Guattari expound, refers to that which has the "capacity to make the elementary communicate with the cosmic: precisely because it effects a dissolution of form that connects the most diverse longitudes and latitudes, the most varied speeds and slowness, which guarantees a continuum by stretching variations far beyond its formal limits."[63] Such a continuum across entities is illustrated by the rhizomic structure, which is defined as "subterranean stem" running on the principles of "connection and heterogeneity."[64] Deleuze and Guattari use the orchid-wasp relationship as an example of the rhizomic becoming.[65] It provides a model of imagination more profound and expansive than the conventional idea of pity for it allows us to feel for the death throes of a lower creature, such as a rat.[66] It fits in the bottom-down scale of concerns expressed by the framing device of the novel. As Derek P. McCormack describes, the molecular "designat[e]s a zone of composition, decomposition, and recomposition working below the perceptual and affective thresholds of organized forms and subjective territories."[67] In Chang's novel, the emblem of the brothers with ants suitably serves as a visual approximation of the Deleuzean insight, with the ants serving as an apt proxy of deterritorialization and reterritorialization. This molecular model can be deployed as a way for the brothers to feel for and become each other and for the ants to feel for and become them, and for the reader to vicariously participate in the strife.

The first chapter of the novel unfolds with the aftermath of the battle as fallen camphor leaves shine like cicada's wings. Ants crawled over the end of Walis's nose which was stained with gunpowder, blood, and dust. All would be decomposed to provide nutrients to other life forms.[68] The reference to cicada suggests metamorphosis, the transformation of life among its various forms, which is in a parallel to Deleuze and Guattari's idea of "involution" that "runs its own line 'between' the terms in play and beneath assignable relations," and an rhizomic "alliance" of life force.[69] Body fluids (sweat, blood, breast milk, etc.), tree saps, and gunpower were mingled. All these intertwining and intermeshing announce the spirit of the land that requires us to acknowledge its minute and evolving physical makeup. There is an even more shocking pairing set in the backdrop of the headhunting ritual: the cooking of human flesh and the distillation of camphor crystal. This series of parallels is again set within a larger pairing: the facial tattoos of the Atayal and the grained barks of the camphor tree, suggesting a granular, immanent, and cosmic correspondence between the human and the nonhuman.

The novel ends with the two brothers maiming each other and lying paralyzed beside two camphor trees.[70] It is almost Greek tragedy-like, as they seemed to recognize each other at their last breath, with their wives killing each other in the background, and no one left to tell their stories. Decades afterward, a hiker came near the site, appreciating the fragrance of camphor.[71] The mountains are indeed immersed in the camphor scent which is issued in strife both physically as the tree sap needs a cut on the barks to emanate scent and spiritually as the site marks bloody wrangles. Chang is judicious at this juncture to bring forth the ants again. The frame of ants broadens and deepens the rude contention to the two-fold vicissitude of geopolitics and biosphere. The framing device can also be read as suggesting a Deleuzean "refrain" for it repeats and returns with a twist on psyche and history.[72] The ants' refrain requires to be considered together with headhunting, cannibalism, and distillation of camphor, as instances of deterritorial-reterritorial molecular out-reaching.

The novel is thus bookended with slow-motion, zoomed-in accounts of the battleground with humans, ants, and all other creatures embroiled in the aftermath of the drastic fight, itself a repetition of a major fight almost twenty years ago. The ants' colonies were disturbed by the havoc just as the human population was devastated by the wrangle. The ants were trying to find the trace of their queen and go back to their nest. The human settlements on both sides fared far worse for they had razed each other to the ground.

The parallel between the human strife and the ants' struggle finds a classical allusion to a passage in Virgil's *Georgics* on "The Fighting Swarm":

> When a quarrel breaks out between two rival kings
> And the bees come out to battle, swarm against swarm,
> . . . .
> Where the fighting is thickest, the kingly champions spread
> Their glittering wings, great souls in little bodies,
> Determined, both of them, to fight to the end—
> . . . .
> Yet their high passions, their desperate encounters
> Are easily quietened by flinging a handful of dust.
> Once you have stopped the fighting and recalled both captains,
> Destroy the weaker one—weakness brings waste—
> And leave the stronger to reign unchallenged in the hive. (4.65–90)[73]

Virgil's manual on agriculture is written at a time of political turmoil. Therefore, the fight between the two bee colonies can be rightfully regarded as an allegory for human warfare. Virgil recommends the beekeeper to restrain the two bee "generals" by tossing a little dust on them and to keep only the stronger one to prevent waste of resources. The irony appears in sharp relief when we realize there would be no "stronger" victor emerging from this fierce struggle and what's being devastated includes camphor forests and the whole environment. The passage from Virgil allows us to see Chang's insight in the choice of *Crematogasters*, rather than the bees, as the foil to humans. *Crematogasters* are highly territorial social creatures. Farmers often use them to disperse other pests on their crops. Ants are also scavengers which help the process of decomposition. They seem to exist at the lower end of the food chain but their significance is vital. Interestingly, in the novel, the Han Chinese settlers were considered by the Atayal to be so numerous or laboring away on the camphor trunk like the ants.[74] Only the reader is vouchsafed with a potential to gain a perspective beyond and beneath the clash between Han Chinese and the Atayal.

CONCLUSION

Due to its importance in the history of Taiwan, camphor has been employed as an apt figure for Taiwan as in Shu-ching Shih's *A Man Who Has Been through Three Ages*, in which the camphor tree speaks at the end of the novel to foreshadow the irony that the camphor factory was in the vicinity of the Bureau of Monopoly where the 228 Incidence (1947) took place.[75] Shih casts the camphor tree as a witness to the three generations of Han Chinese inhabitants in Taiwan. I consider Chang's novel a successful contrapuntal supplement to Shih's in pairing the Atayal with the Han Chinese settlers in their endeavor to survive the tidal waves of imperial extraction. In the end, this chapter has illustrated how Chang's novel enables the reader to experience molecularly with all the parties, humans and nonhumans, caught up in the contentious course of history with Taiwan and the camphor tree.

NOTES

1. Lin Yi-hung and Hsu Yu-chuen, *Explore the Kingdom of Camphor: Special Exhibition of Taiwan Camphor Industry* (Taipei: National Taiwan Museum, 2010). For more information, see Man-hung Lin, *Tea, Sugar and Camphor and Social-Economic Development of Taiwan, 1860–1895* (Taipei: Linking Books, 1997).

2. Kao Jun-honn, *The Chronicle of Traversing: The War over the Forest of Taiwan, Empire and Image* (New Taipei City: Walkers Cultural Enterprise, 2017), book blurb, 16.

3. Gilles Deleuze and Felix Guattari, *A Thousand Plateaus: Capitalism and Schizophrenia*, trans. Brian Massumi (Minnesota: University of Minnesota Press, 1978), 308–309.

4. Arjun Appadurai, *The Social Life of Things: Commodities in Cultural Perspective* (Cambridge: Cambridge University Press, 2013), 5.

5. Emma Jinhua Teng, *Taiwan's Imagined Geography: Chinese Colonial Travel Writing and Pictures, 1683–1895* (Cambridge, MA: Harvard University Press, 2004), 2.

6. Teng, *Imagined*, 27.

7. Quoted in Teng, *Imagined*, 26.

8. Teng, *Imagined*, 123.

9. Teng, *Imagined*, 139.

10. Teng, *Imagined*, 27.

11. Teng, *Imagined*, 213.

12. Ronald C. Po, "The Camphor War of 1868: Anglo-Chinese Relations and Imperial Realignments within East Asia," *The English Historical Review* 135, no. 577 (December 2020): 1467.

13. R. A. Donkin. *Dragon's Brain Perfume: An Historical Geography of Camphor* (Leiden: Brill, 1999), 228.

14. Donkin, *Dragon's Brain*, 231.

15. Donkin, *Dragon's Brain*, 9–10.

16. Antonio C. Tavares, "Crystals from the Savage Forest: Imperialism and Capitalism in the Taiwan Camphor Industry, 1800–1945," PhD diss. (University of Princeton, 2004): 9–10.

17. Donkin, *Dragon's Brain*, 37.

18. Rebecca Altman, "The Myth of Historical Bio-based Plastics," *Science* 373, no. 6550 (2 July 2021): 48.

19. Tavares, "Crystals," 4.

20. Tavares, "Crystals," 107, n. 1.

21. Donkin, *Dragon's Brain*, 58–59.

22. Yu Chih Chieh, *Traverse Taiwan: On the Phytogeographical Origin of Montane Plants in Taiwan* (Taipei: Spring Hill Publishing, 2023), 159.

23. Donkin, *Dragon's Brain*, 219.

24. James W. Davidson, *The Island of Formosa, Past and Present: History, People, Resources and Commercial Prospects. Tea, Camphor, Sugar, Gold, Sulphur, Economical Plants, and Other Productions* (London: Macmillan, 1903), 398.

25. Yu, *Traverse*, 152–157.

26. Tavares, "Crystals": 6–7.

27. Tavares, "Crystals": 7.

28. Tavares, "Crystals": 7–8, n. 16

29. Tavares, "Crystals": 109.

30. Walis Nokan, "Telephone Receiver," *United Daily Literary Supplement* (15 June 2018), https://paper.udn.com/udnpaper/PIC0004/327983/web/.

31. Po, "Camphor": 1483.

32. Pei-chen Li, "Institutional Change and Commercial Interests: Competitive Strategies in Taiwan's Camphor Trade, 1850–1868," *Xin Shixue* 30, no. 1 (March 2019): 53.

33. Ying-min Chang, *Blood Camphor* (Taoyuan: Taoyuan City Library, 2017), 91–92. Quotations from the novel are my own translation.

34. Tavares, "Crystals": 9–10.

35. Chang, *Blood*, 25.

36. See *Encyclopedia of Taiwan*, https://nrch.culture.tw/twpedia.aspx?id=5045.

37. Chang, *Blood*, 30.

38. Chang, *Blood*, 90.

39. Chang, *Blood*, 192.

40. Chang, *Blood*, 69.

41. Chang, *Blood*, 104.

42. Chang, *Blood*, 118, 150.

43. Chang, *Blood*, 214.

44. Chang, *Blood*, 126.

45. Chang, *Blood*, 92, 133.

46. Chang, *Blood*, 192.

47. Chang, *Blood*, 35.

48. Chang, *Blood*, 104.

49. Chang, *Blood*, 66.

50. Chang, *Blood*, 59.

51. Chang, *Blood*, 55.

52. Chang, *Blood*, 195.

53. Chang, *Blood*, 182.

54. Quoted in Scott Simon, "Politics and Headhunting among the Formosan Sejiq: Ethnohistorical Perspectives," *Oceania* 82 (2012): 164.

55. Quoted in Simon, "Politics": 165.

56. Simon, "Politics": 165.

57. Quoted in Simon, "Politics": 171.

58. Simon, "Politics": 177.

59. Simon, "Politics": 164–65.

60. Chang, *Blood*, 120.

61. Chang, *Blood*, 120.

62. Chang, *Blood*, 121.

63. Deleuze and Guattari, *Thousand*, 308–309.

64. Deleuze and Guattari, *Thousand*, 6, 7.

65. Deleuze and Guattari, *Thousand*, 10, 12, 294.

66. Deleuze and Guattari, *Thousand*, 258.

67. Derek P. McCormack, "Molecular Affects in Human Geographies," *Environment and Planning A: Economy and Space* 39, no. 2 (2007): 368.

68. Chang, *Blood*, 20–22.

69. Deleuze and Guattari, *Thousand*, 239, 25.

70. Chang, *Blood*, 211.

71. Chang, *Blood*, 220.

72. Deleuze and Guattari, *Thousand*, 299.

73. Virgil, The *Georgics*, trans. Robert Wells (Manchester: Carcanet New Press, 1982), 81.

74. Chang, *Blood*, 100, 59.

75. Chu-ching Shih, *A Man Who Has Been through Three Ages* (Taipei: China Time Publishing Co., 2010), 271–273.

BIBLIOGRAPHY

Altman, Rebecca. "The Myth of Historical Bio-based Plastics." *Science* 373, no. 6550 (2 July 2021): 7–49.

Appadurai, Arjun. *The Social Life of Things: Commodities in Cultural Perspective*, Cambridge: Cambridge University Press, 2013.

Chang, Ying-min (張英珉). *Blood Camphor* (血樟腦). Taoyuan: Taoyuan City Library, 2017.

"Ching Kuang Fu" (金廣福). *Encyclopedia of Taiwan* (臺灣大百科), https://nrch.culture.tw/twpedia.aspx?id=5045.

Deleuze, Gilles, and Felix Guattari. *A Thousand Plateaus: Capitalism and Schizophrenia*. Translated by Brian Massumi. Minneapolis: University of Minnesota Press, 1978.

Davidson, James W. *The Island of Formosa, Past and Present: History, People, Resources and Commercial Prospects. Tea, Camphor, Sugar, Gold, Sulphur, Economical Plants, and Other Productions.* London: Macmillan, 1903.

Donkin, R. A. Dragon's Brain Perfume: An Historical Geography of Camphor, Leiden: Brill, 1999.

Kao, Jun-honn (高俊宏). *The Chronicle of Traversing: The War over the Forest of Taiwan, Empire and Image* (橫斷記：臺灣山林戰爭、帝國與影像). New Taipei City: Walkers Cultural Enterprise, 2017.

Li, Pei-chen (李佩蓁). "Institutional Change and Commercial Interests: Competitive Strategies in Taiwan's Camphor Trade, 1850–1868." *Xin Shixue* 30, no. 1 (March 2019): 1–53.

Lin, Man-hung (林滿紅). *Tea, Sugar and Camphor and Social-Economic Development of Taiwan, 1860–1895* (茶、糖、樟腦業與臺灣之社會經濟變遷 1860–1895). Taipei: Linking Books, 1997.

Lin, Yi-hung, and Yu-chuen Hsu (林一宏、許毓純). *Explore The Kingdom of Camphor. Special Exhibition of Taiwan Camphor Industry* (探索樟腦王國特展). National Taiwan Museum (國立臺灣博物館), 2010.

McCormack, Derek P. "Molecular Affects in Human Geographies." *Environment and Planning A: Economy and Space* 39, no. 2 (2007): 359–377.

Po, Ronald C. "The Camphor War of 1868: Anglo-Chinese Relations and Imperial Realignments within East Asia." *The English Historical Review*, vol. 135, no. 577 (December 2020): 1461–1487. https://doi.org/10.1093/ehr/ceaa313.

Roy, Toulouse Antonin. "'The Camphor Question Is in Reality the Savage Question': Indigenous Pacification and the Transition to Capitalism in the Taiwan Borderlands (1895–1915)." *Critical Historical Studies* 6, no. 1 (2019): 125–158.

Shih, Chu-ching (施叔青). *A Man Who Has Been through Three Ages* (三世人). Taipei: China Time Publishing Co., 2010.

Simon, Scott. "Politics and Headhunting among the Formosan Sejiq: Ethnohistorical Perspectives." *Oceania* 82 (2012): 164–185.

Tavares, Antonio C. "Crystals from the Savage Forest: Imperialism and Capitalism in the Taiwan Camphor Industry, 1800–1945." PhD diss. University of Princeton, 2004.

Teng, Emma Jinhua. *Taiwan's Imagined Geography: Chinese Colonial Travel Writing and Pictures, 1683–1895*. Cambridge, MA: Harvard University Press, 2004.

Virgil. *The Georgics*. Translated by Robert Wells. Manchester: Carcanet New Press, 1982.

Walis Nokan (瓦歷斯· 諾幹). "Telephone Receiver" (機筒). *United Daily Literary Supplement* (15 June 2018). https://paper.udn.com/udnpaper/PIC0004/327983/web/.

Yu, Chih chieh (游旨价). *Traverse Taiwan: On the Phytogeographical Origin of Montane Plants in Taiwan* (橫斷臺灣：追尋臺灣高山植物地理起源). Taipei: Spring Hill Publishing, 2023.

Chapter 6

Beyond Cold War Afforestation

Trans-scalar Imaginary of Casuarinas

Wei-bon Wu

This chapter explores the ecological, social, and cultural values of casuarina trees in the post-Cold-War Jinmen (alternatively known as Kinmen or Quemoy).[1] The *Casuarina* spp.[2] were introduced to Jinmen for their capacity to adapt to adverse climatic and edaphic conditions. Casuarinas are the most ubiquitous and observable features on the landscape other than man-made structures. During the Cold War era, they flourished in turbulent times. Under the attentive care of tree planters, the saplings grew in the hope that they would ameliorate the island's livability. In this chapter, I trace the casuarina's historical trajectories from their cultivation to the present in order to reflect on their manifold entanglements with the landscape as well as the mindscapes of the islanders. From individual saplings, assemblages, to forest patches, the scales of casuarinas underwent dramatic transformations within decades. At the current stage, the casuarina's values are being questioned as they are no longer deemed relevant to the island's post-militarization development.

THE HISTORICAL CONTEXT

The casuarina's history in Jinmen has intertwined with the history of anthropogenic disturbances. After the 1949 Battle of Guningtou, the number of Nationalist troops grew rapidly across the Jinmen archipelago. The troops' primary mission was to fortify the front line in order to secure Taiwan. The number of soldiers soon surpassed that of civilians. To strengthen the border island's environmental resilience and sustainability, the troops and the civilians were ordered to plant trees. Millions of casuarina saplings became a green task force. This vegetal fortress was expected to carry out the following missions: to reduce soil erosion, to suppress sand drift, and to shield and camouflage military facilities. A dozen years later, assemblages of casuarinas effected a large-scale landscape transformation. An abundance of newly naturalized trees—each of them could be linked to an individual soldier planter who provided attentive care to the sapling—engendered impacts on the island's material and cultural aspects. For the islanders, the omnipresent sylvan on the militarized landscape pervaded in various dimensions on their mindscapes.

Under the watchful care of soldiers and civilians of Jinmen, a large-scale reforestation program transformed the once sandy and rocky island during the Cold War era. Pertaining to the era, Jun-Yao Wu's literary works[3] provide multifocal and multidimensional autoethnographic accounts of Jinmen, enabling us to see the casuarina's relationships with the islanders across space and time. As the Cold War raged, casuarinas spread steadily over the island, and the woody biomes became more entangled with the local residents. Reflecting the greener battlespace, Wu's writings also feature casuarinas as a recurring motif. As a matter of fact, he grew up surrounded by the burgeoning ubiquity of sylvan assemblages, the large-scale biomass created by the cooperative military-civilian mobilizations.[4]

How can we fathom the vast quantities of data about casuarina trees that disclose both the human and nonhuman histories and the emotional ties between humans and plants? To explore the multiple dimensions of casuarina atlas, I will apply the frameworks promoted by Scott Slovic and Anna Tsing—"trans-scalar thinking" and "nonscalability theory" respectively. In his essay, "Liquid Scale: Trans-Scalar Thinking and the Perception of Water," Slovic describes what "trans-scalar thinking" entails: "This means the ability to move cognitively across spatial and temporal scales: for instance, from small to large, from the individual to the collective, and from the present to the future."[5] Comparable to Slovic's multidimensional interpretation of scales, in "On Nonscalability: The Living World Is Not Amenable to Precision-Nested Scales," Anna Tsing proposes a method to call into question the trend or "quality called scalability, that is, the ability to expand without rethinking basic elements. Scalability is, indeed, a triumph of precision design, not just in computers but in business, development, the 'conquest' of nature, and, more generally, world making."[6] As a critique of the domineering mass production of scalable materials, Tsing puts forward "a nonscalability theory that pays attention to the mounting pile of ruins that scalability leaves behind."[7] Essential to Tsing's idea of "nonscalability" is a transformative relationship, which draws more attention to the change in qualities rather than uncontaminated quantitative growth and exponential expansion devoid of diversified processes.

To examine the casuarina's trans-scalar effects and transformative impacts on the island of Jinmen, Yi-fu Tuan's approach to and interpretation of landscape also provides profound insights in my reading of temporal and spatial scales involved in Jun-Yao Wu's literary oeuvre. In *Topophilia: A Study of Environmental Perceptions, Attitudes, and Values*, Tuan writes: "Literature rather than social surveys provides us with the detailed and finely shaded information on how human individuals perceive their worlds."[8] In Wu's writings, memories of the good old days are preserved primarily through his perceptions as a child before moving to Taiwan as a teenager. In the turbulent years of the Cold War era, casuarinas that grew in profusion served social functions. For Wu, they also provided a sense of security and emotional anchorage. With the passage of time, however, the end of the Cold War ushered in new forms of economy in the 1990s. With new ideas to pursue economic growth, the landscape was reshaped, and the omnipresent casuarina assemblages were overlooked in the process. The outdated military existence was no longer relevant; the post-militarization landscape had to accommodate new modes of life. The vestiges of military fortifications that sprawled all over the island were progressively superseded by development projects. Jinmen experienced social and cultural changes and embraced post-Cold War era economies.

Still, some vital issues need to be addressed. What social-cultural-ecological values can still be brought to light through the assemblages of casuarinas that have survived the Cold War island of Jinmen? What new terrain can we envisage through the casuarina ecosystems that stand as vegetal reminders of the military past? The trajectory of this chapter will be concluded with Caitlin DeSilvey's idea of "curated decay."[9] DeSilvey proposes an unorthodox approach to acknowledging the inevitable change and loss of ecologically and culturally significant spaces by giving prominence to the transformative processes of decay. Through reading Jun-Yao Wu's stories, this chapter explores how the casuarina can be a focal point in discussing complex issues in relation to memory and vegetation. In Wu's autoethnographic narratives, the casuarina induces trans-scalar imaginary, which innerves our perceptive dimensions to navigate temporal, spatial, and quantitative scales.

In this chapter, by contextualizing narratives from Wu's *Jinmen* (金門, a collection of Jinmen-inspired short prose pieces), *Three Tree Friends* (三位樹朋友, an illustrated children's book), and *Maps of Heat* (熱地圖, a body of 66 prose pieces that provides clues about the author's mental atlas), I propose that the survival of casuarina trees is not only capable of triggering trans-scalar imaginary and evoking narrative empathy, but is also essential to the ongoing rehabilitation of the human ruins in the Cold War era. Casuarinas play a dual role in the post-militarization era, transforming the environment and caring for the land of the island. I will evince how they replenish the island landscape with their restorative capacity. Wu's sylvan narrative reveals that for an islander who has had a close connection with the casuarina community, memories of individual trees can conjure up the ambiance of home, which, as Tuan defines, is "the locale of human warmth and material sustenance, moral probity and spiritual comfort."[10] In Wu's stories, a casuarina is a portal to another place, both real and imagined, a living reminder of home even when he is elsewhere. In the following section, I will trace back to the early stages of the plant's cultivation history and relationships with the island of Jinmen, which measures about 150 square kilometers.

BEFORE AND AFTER 1949: NUMBERING THE CASUARINAS

The first page of *Three Tree Friends* pictures the storyteller (Jun-Yao Wu) as a child. Three individual trees stand in front of him: "I have three tree friends. They are the most common tree species [on Jinmen]: casuarina, banyan, and acacia."[11] Throughout the illustrated book, interpersonal and interspecies relationships center around casuarinas, which give support to other ecosystems. For a child growing up in the 1960s in Jinmen, the encounters with casuarina were most natural since they were the most widespread species and greatest in numbers. At a rapid pace, casuarinas rescaled the island's vegetation. Despite their omnipresence, they were newcomers on the island. Since the 1950s, casuarinas progressively transformed the largely treeless landscape of Jinmen and ushered in a new period of greenness. Before 1949, the island had long been subject to the scarcity of vegetation. In *The Chronicles of the Open Water*,[12] Shou Hong, a scholar of Confucian hermeneutics in the Ming Dynasty, provided the following statement: "Agriculture is the foundation of people's sustenance. When farmers fail to cultivate land, they lose their livelihoods. Jinmen's arable land is sparse, and the soils are infertile. Moreover, owing to the gusty winds and drifting sand, farmers find themselves in an unsustainable position."[13]

Almost four centuries after the publication of Shou Hong's historical accounts, Jinmen still endured sparse vegetation and the effects of extensive deforestation. The year 1949 marked a transition and a breakpoint in the history of Jinmen. The year of turmoil transformed the relatively untroubled island, an obscure dot on the world map, into a military flashpoint. Prior to the geopolitical conflict, the islanders of Jinmen shared close social and economic ties with Xiamen (and other ports in Fujian province). In 1949, the island's historical continuity was severed by military dominance and turbulence. In the following decades, the marginal island's history was divided by force under the sway of geopolitics. Since 1949—the year in which the ad hoc militarization[14] was enforced—the island's history was divided into phases according to its role dictated by the need for Taiwan's defense. Even plants were taken as a part of military mobilization.

From the 1950s onward, Jinmen had been transformed by the extensively cultivated casuarinas, which constitute part of the ad hoc fortifications. Parallel to the inconspicuous underground maze of military tunnels, casuarinas became distinctly visible features on the island's surface. An article published by *Taiwan Today* in 1960 opened by saying: "Trees planted on a hillside to prevent erosion may grow faster if they are given good care. This same idea may be applied to Jinmen."[15] As a project of militarized development, the burgeoning presence of casuarinas emblemized the military power to take "good care" of the island and transform it into a frontline fortress. Within a decade, the army's power to mobilize the soldiers and civilians gained efficacy. Straight roads lined with casuarinas crisscrossed the island. The trees stood guard along the emergent roads, marking the boundaries, and offering shade to soldiers and civilians. The article continues:

> Jinmen's sandy hills are no longer barren and denuded as they were a few years ago. . . . Afforestation has won support from both the civilian government and the army headquarters, particularly the latter because the army is interested in tree planting for camouflage. A total of 2,107,735 seedlings of some nine kinds of trees, mostly casuarina, have been planted all over the island.[16]

The description that the newly planted trees or the island were "given good care" is an understatement. The effects of militarization severely disrupted the islanders' pre-1949 way of life. Despite the military tensions and artillery shelling, a tremendous scale of casuarinas thrived during the sea change of geopolitical disturbances. Casuarinas reinvented the island. Along with the newly paved roads that connected military facilities, the introduced plants reconstructed and rescaled the vegetation of Jinmen. The militarized island was transformed into a green sentinel, extensively camouflaged by casuarinas. A 1976 *New York Times* article entitled "Still Waiting for War" gave the following depiction:

> After 1949, we wanted to "greenize" the whole island, says one of the omnipresent escort officers. "We have planted 78 million Australian pine trees to cover the barracks and garrison." The figure may strain credibility, but his remark illustrates both the underlying poverty of the islands and their almost□total military function. On both island groups, the military outnumber the civilian population. Quemoy is home to about 60,000 farmers, fishermen and merchants.[17]

In the correspondent's eyes, the number of casuarinas planted prior to 1976 may stretch the bounds of credulity. However, it implies the extraordinary scale of military-civilian engagement in the greening operation. The anthropogenic magnitude of "green infrastructure" was the result of the collective work and intensive "good care" provided by a significant number of soldiers and civilians—including a great number of mobilized farmers, fishermen, and merchants.

CONNECTING SINGULARITIES: DIALOGUE BETWEEN CASUARINAS AND THE ISLAND

The casuarina afforestation pervaded Jinmen on a comprehensive scale in a matter of years. How did the vast number of casuarinas affect the islanders' day-to-day war zone existence, and how do we comprehend the dialogue between the islanders and the plants? The greening operation involved the combined force of individual islander's scrupulous care. The threads of life's storylines grew vastly numerous and entangled, but a singular focal point may provide a framework for identifying intricate connections in a large-scale context. In his article "Language Matters: Environmental Controversy and the Quest for Common Ground," Scott Slovic says:

> Another key aspect of storytelling is the conventional focus of stories on small-scale phenomena, highlighting small groups of characters or individual characters, even when the point is to express the condition or experience of larger classes of people or animals (or other abstract phenomena). Our human ability to pay attention and feel empathy is extremely limited—we are prone to become insensitive to phenomena that exceed our capacity to care.[18]

As the numbers involved in a phenomenon increase and proliferate, we are at the risk of losing a focal point and becoming inattentive. Slovic's countermeasure is to draw our attention to the singularity effect, which gives anchorage as we process a plethora of information, grand narratives, or compounding problems. To comprehend the immeasurably entangled threads of life, singularity effect creates a pathway to narrative empathy, in which emotional anchors and values take shape. To comprehend the complex casuarina effect, I will navigate the temporal and spatial scales with a focus on Wu's storytelling through his encounters with individual casuarina trees.

During the Cold War era, while precarity defined the islanders' day-to-day existence, casuarinas appeared to offer shelter, material support, and emotional anchorage. In the preface to *Jinmen*, Wu states: "Each generation is divided and shaped by their historical periods. Our generation lived between billowing sea and swaying trees."[19] Wu's family dwelled in a traditional fishing village called Xiguo Hill. The villagers depended on marine catches for

sustenance and traditional fishing practices were a precarious way of life due to the unstable weather conditions.

Having experienced the fishing village's winter windstorms and summer typhoons, a sense of security acquired from the casuarina is deeply engrained in Wu's childhood memories. Even after spending decades living in Taipei, this sense of emotional anchorage still holds when he witnessed his native village falling into a state of ruination. In a piece of prose entitled "Dilapidated," he writes: "My native village 'Xiguo Hill' is decrepit."[20] He continues to say: "You can imagine it as a big balloon tethered to the swaying branches of a casuarina, swaying in the rumbling sound of crashing waves. . . . "[21] Here, a poignant moment is exposed and mingled with imagination. As an adult who returned to his childhood sanctuary, Wu found it in an irreversible process of decay. In contrast to his joyful childhood memories, where the casuarina trees were his companion species and steadfast friends, his adulthood has been haunted by a brooding sense of loss. At such a moment, a small-scale singularity effect is communicated through the casuarina: an emotional anchorage and an imaginary constituent of home.

Through Wu's stories, the readers are transported across spatial and temporal scales via his autoethnographic narrative and empirical data. Within decades, the patches formed by tens of millions of casuarinas stretch beyond human scope and power of perception. However, a child's standpoint may offer a sense of clarity. In addition, a child's image is likely to evoke memories of poignancy, even a strong sense of loss. Slovic writes:

> If understanding the singularity effect helps us to appreciate the power of individualized narratives in galvanizing attention and prompting at least short-lived responses to a crisis, the lesson of the Aylan Kurdi story and other similar stories. . . . is that we really need multidimensional communication strategies in order to adequately convey information and spur appropriate public and governmental responses.[22]

Slovic explores how we are more likely to feel compassion for a single victim than for a large group of anonymous victims as a result of emotional limitations that prevent people from fully grasping the extent of a large-scale suffering. He illustrates this point with the example of vast numbers of refugees who fled the Syrian Civil War and risked their lives to seek safety. Aylan Kurdi's[23] image stirred up an awakening sense of loss in the Kurdish diaspora. A lone image of the child's demise was a poignant microcosm of the large-scale tragic event that had uprooted millions of refugees.

In the context of environmental issues, attentive observation of single individuals rather than a surplus of unidentified ones can also be an approach that facilitates our apprehension of the casuarina effect. While bearing in mind that singularity effect is not quantifiable, however, it can be transferred and intensified in the form of accretion. In the following section, I will move between singularities and multifarious scales: from individuals to an aggregate of singularities that are entangled in the atlas of casuarina-connected stories. Relations between singularities are dialogic, accretive, and transformative; the atlas of connectivity invites us to think across scales. In the following section, I intend to prove that the singularity effect can be communicated and propagated through the accretion of transformative relations.

ACCRETION: A TRANS-SCALAR IMAGINARY

In *Three Tree Friends*, casuarina, banyan, and acacia are singled out as the tree species that give structure to Wu's idea of home and community. Transformative relationships between the author and three singular trees took root from an early age. Singularity effect accretes and forms into an organic network: a map of relationships which extends and transforms as the constituents grow, mature, age, and decay. In this perennial process, ecological and cultural values accumulate. The human and nonhuman values overlap and enmesh in a varied assemblage of maps: an atlas of stories.

Singularity effect gives us an anchorage, from which we navigate the open sea of information and the complexity of relations. As Anna Tsing says in *The Mushroom at the End of the World*: "To listen to and tell a rush of stories is a method."[24] Tsing continues: "To learn anything we must revitalize arts of noticing and include ethnography and natural history. But we have a problem with scale. A rush of stories cannot be neatly summed up. Its scales do not nest neatly; they draw attention to interrupting geographies and tempos."[25] Since the onset of the Cold War, the widely cultivated casuarinas infiltrated Jinmen's daily activities. Casuarinas came to become companions to the islanders' life stories. The introduced species covered and altered the terrain. Over time, the plants became a nonscalable phenomenon in the sense that they carry on transforming and rescaling the island's fabric of relationships. In this section, I aim to revitalize the art of noticing by forming an imaginary atlas of the casuarina's history and mapping their transformative relations with the islanders.

Casuarinas in Wu's stories enable us to imagine a spectrum of spatial and temporal scales. A casuarina is a portal, through which we roam the atlas of relations. The atlas houses layers of co-constituted entities: values and meanings proliferate in the cohort of natural-cultural strata. We forge transformative relationships by weaving our way through the intricate web of connections. It is also interesting to note that in the illustrations of *Three Tree Friends*, some soldiers are portrayed as tree stewards. At the beginning of the book, one soldier is pictured planting a sapling; in the central pages, one soldier waters a young casuarina; a few pages later, one soldier raises his head and looks at a skyward-reaching tree; near the story's end, one soldier prunes a casuarina among two straight rows of lofty casuarinas. Through this process, stories between casuarinas and humans emerge. Dynamic relations evolve, transform, and accrete over time.

During the Cold War period, most troops were only stationed on Jinmen for the duration of their service obligation. Yet, with the collaboration of the tree-planting force, woodlands with varied scales continued to proliferate on the island. Casuarina patches evolved into the island's natural cultural heritage; they served as places of connectivity in Jinmen's militarized society. Once an adventive species, the casuarina was integrated into the island's landscape and social fabric. While casuarinas now have a pervasive presence, their existence can be traced back to the tree-planting operation of a member of the military or militia in the Cold War period. In the early stages of reforestation, the degraded and windblown soils made woodland creation a strenuous task. Tree stewards had to be extremely attentive to the saplings; an affinity was formed between the cultivator and the cultivated. The casuarina became a tangible manifestation of the person's endeavors and, on an individual scale, the steward's memories, as an intimate relationship was established in the process.

To varying degrees, casuarinas shape the (ex-)servicemen and islanders' identities and sense of place. In *Cold War Island: Quemoy on the Front Line*, Michael Szonyi records this period of Jinmen's silviculture history: "Forestation was also tied to the issue of popular mobilization, as militia members as well as soldiers were charged with planting and caring for seedlings. 'In the 1950s, every tree had a little plaque with the name of the person responsible for planting it and looking after it,' recalls one former magistrate."[26] From the statistical figures collected by Szonyi, 42,000 troops and 6,000 guerrillas were stationed on Jinmen in 1954. The figures of troops reached a staggering 100,000[27] in 1959. It is not difficult to imagine that troops at such a large scale could plant tens of millions of trees under the military command. According to the statistics recorded by the Jinmen County Forestry Bureau, the total amount of trees planted between 1955 and 1997 reached 76,466,843 (among which, 19,186,904 were casuarinas). In a sense, the militarized island of Jinmen was a botanic garden on a gigantic scale.

Emotion-wise, these statistical figures probably signify nothing in the human perceptual process. Vast quantities have little bearing on emotional perception and empirical experience. On the contrary, exponential growth of numbers is likely to increase the risk of psychic numbing. In *Numbers and Nerves*, Slovic and his psychologist father Paul Slovic write:

> Instead of numbing audiences with blizzards of nerveless information, skilled communicators can navigate accurately and vividly between large-scale phenomena and small-scale illustrations, between the remote and the proximate. The purpose of such discourse, of course, is to foster audience *attachment* to the issues and phenomena being described, and crucial to the formation of attachment are *proximity* and *scale*.[28]

Feelings and emotions do not intensify in line with the escalation of numbers: "Even the most mathematically gifted human beings are psychologically limited when it comes to attaching feeling to numerical information."[29] Feelings and emotions revolve around a focal point, an anchorage. Attachment emerges from a profound visceral process of sensing proximity. Trans-scalar imagination is an enabler of such transformative relations. A sense of longing and belonging grows from this resonant experience of affinity: being close to a locale, a house, an object, a tree. As Tuan says: "Many places, profoundly significant to particular individuals and groups, have little visual prominence. They are known viscerally, as it were, and not through the discerning eye or mind. A function of literary art is to give visibility to intimate experiences, including those of place."[30] Literary works give us a focus on forming a visceral sense of place, an anchorage in a sea of sensory data. Tuan continues, "Literary art can illuminate the inconspicuous fields of human *care*."[31] Home is a place where intimate experiences occur. Traditionally, at the heart of home is hearth, which illuminates and reifies the idea of home. During the Cold War period, the abundant leaf litter and cones of casuarina trees were the most common and ideal hearth firewood on Jinmen. The food cooked over the hearth fire had fed every islander. This communally but individually experienced intimate relationship was brought into being by the cultivators' attentive care of casuarina saplings: the accretion from one to millions more.

For any child who grew up on the Cold War island, the aroma of food cooked over casuarina-wood fire was an indelible experience. *Three Tree Friends* also portrays a kitchen scene in which two children are given a treat by a mother. At one corner is the hearth; the other corner features a table with steamed cakes on it. Using casuarina cones as tools, the children decorate the cakes with appetizing hues of red. This scene captures a heartwarming family moment. It is reminiscent of a time when casuarinas were cared for and their social-cultural values were illuminated by transformative relations.

ARTS OF NOTICING: VALUES OF DECAY

In the early stages of intensive militarization, casuarina afforestation was implemented on Jinmen in order to regenerate green space. On a micro-scale, the casuarina's nitrogen-fixing capacity cures the depleted soils (the chemical transformation occurs without being perceptible to the human eye). On a macro-scale, their presence created forest patches that formed squadrons of green forces all over the island. Since the 1990s, Jinmen entered a new period of demilitarization. The military had withdrawn most of its troops from the fortress island. Former military facilities that are scattered over the map of Jinmen are subject to progressive ruination and other forms of destruction—nearly all the abandoned sites have been dominated by advancing casuarinas.

The military ruins, undisturbed by humans, are growing feral in the post-militarization era. Several designated ex-military installations are artificially preserved as historical heritage. The abandoned sites encapsulate memories of the military past. Some chosen ex-military sites are targets of attentive care and preservation while most are left to decay. In *Curated Decay: Heritage beyond Saving*, Caitlin DeSilvey remarks: "We live in a world dense with things left behind by those who came before us, but we only single out some of these things for our attention and care. We ask certain buildings, objects, and landscapes to function as mnemonic devices, to remember the pasts that produced them, and to make these pasts available for our contemplation and concern."[32] DeSilvey presents an unconventional practice of recognizing the value of materials as they decay: a careful and deliberate way of letting things be while embracing and managing a sense of loss as a spontaneous response to the irreversible. Evanescence of matter is an essential feature of the transformational landscape; the values of decaying matter are released through our attentiveness to the transient intermediate state.

In *Cold War Island*, Szonyi records the following remark from "two American observers": "On Jinmen itself there are soldiers, more soldiers than civilians. They run the island. They are responsible for its improvements, its roads, its trees, its wells and system of irrigation, its schools . . . its safety."[33] Planted to guard the perimeter of military bases and newly paved roads, casuarinas were part of the provisional defense infrastructure in the process of creating a militarized island; they became the accretive organic matter that formed the defense system. From multiple empirical data, it is evident that casuarinas were a constituent part of the island's material past. Thanks to their fast-growing capacity, casuarinas had ad hoc tasks to accomplish during the Cold War period. In the post-militarization social and economic context, however, as with a great number of ex-military sites, the casuarina's presence is frequently dismissed as irrelevant in the present-day Jinmen, or as a discordant element in the new urban landscape.

How do we rethink the casuarina in the island's current sociocultural setting? One of the purposes of this chapter is to evoke memories associated with the plants' military past and attest that they carry on conducting natural-cultural transformations in new frontiers. The casuarina was introduced to Jinmen as an environmental amelioration practice that remediated the island's effects of deforestation. The plants also created spaces of dialogue between humans and the green landscape. In the post-militarization era, despite the substantial loss of woodlands and fragmentation of the landscape, casuarinas persist in greening the island, particularly in ex-military sites or places less disturbed by human activities. The ex-military botanic garden continues to revitalize, re-naturalize, and rewild the island.

In a piece titled "The Fragmentary Hill Trail," Wu writes: "Few people know the hill trail that connects Xiguo Hill to Blue Sky Theater except for a few soldiers and villagers."[34] Wu continues: "Many years later, the hill trail is reduced to ruins . . . it is covered in vines, along with assemblages of casuarinas and acacias. Together they form into an entangled darkness."[35] As a pioneer species, the casuarina had been the focus of attention during the Cold War period; tree planting embodied the connection between the military and the civilian, the entanglement of a militarized society and an eco-cultural system. After decades of accretion, their extensive material scale ensured countless possibilities of germination in every nook and cranny throughout the island. The island has been transformed into a seed bank of casuarinas. In the post-militarization period, casuarinas still proliferate in the crevices among piles of ruins and vacancies left behind by soldiers. Despite the tides of capital investments, casuarinas have grown autonomously devoid of human care. When the man-made constructions were abandoned and superstructures have begun to disintegrate, an advanced force of casuarinas began to fill in the gaps before long. The plants that used to mark the boundaries of military facilities and the roadside now connote the collapse of boundaries. Their enduring presence is a continuum of the dialogue between the plants and the ruins where entangled casuarinas keep on thriving.

As a material manifestation of militarization and a form of living heritage, casuarinas register the island's historical process and the disruptive period of war zone administration. On an island strewn with overgrown ruins, the living history of casuarinas provides an anchorage for memories. In *Curated Decay,* DeSilvey writes: "Cultural remembering proceeds not through reflection on a static memorial remnant but through a process that slowly pulls the remnant into other ecologies and expressions of value, accommodating simultaneous resonances of death and rebirth, loss and renewal."[36] Regarding the casuarina's history on Jinmen, "a rush of stories" created by Wu is instrumental in instigating a process of reminiscence and releasing value. The emotional expression of his writings guides the reader's attention to the stories where the dialogue between humans and trees is attainable. Wu writes in a short prose titled "Planting Flowers on My Disability": "Candor is integral to my writing of prose pieces about family affection as well as big history and grand narrative on Jinmen. The first thing to learn about writing, I think, is to care for your own feelings . . . that provides a solid foundation to my works."[37] The sensibilities conveyed through Wu's literary art prompt the reader to explore the transformative relationships between a place, a child, and individual trees in the context of militarization. Traversing the porous boundaries of emotions leads the way toward experiencing the human-plant entanglements; it also ramps up the arts of noticing the post-militarization ruins, transforming them into living landscapes "available for our contemplation and concern" (Wu 92).

Wu's sylvan narrative can be construed as a process of mapping the landscape of emotions and reconfiguring a mental atlas in response to the changing environment. The excessively militarized society had transformed the island on a grand scale. Green patches of casuarinas continue to have the freedom to thrive. In the post-militarization period, casuarinas have been viewed as a hindrance to land development projects. Yet, the pioneer species carries on occupying the liminal spaces and thriving in the military ruins. Social-cultural-ecological values are released through the transformative ruination. This process navigates the space-time entanglements, folded in the atlas of stories encompassing casuarinas. It invites us to exercise what Slovic calls "trans-scalar imaginary" and what Tsing calls "arts of noticing." The plant-human spectrum is a rush of interwoven stories that has evolved into a transferable living heritage; and a spectrum of information filtered through literary works reinforces anchorage to human emotions. As Tuan suggests: "Literary art draws attention to areas of experience that we may otherwise fail to notice."[38] In a short prose titled "Maps of Heat," Wu writes: "I pick up a fallen casuarina branchlet and rub it between my fingers. It crumbles into fragments. I imagine these fragments transform into fire. They conjure up my old map."[39] This passage not only reflects singularity effect but also resonates with DeSilvey's message in *Curated Decay*: "Our perception of the part, rather than the whole, opens up a space that invites speculation and connection."[40] Evolved from the collective cultivation during the Cold War period, casuarinas have integrated into the micro/macro local ecologies and the island's socio-cultural fabrics. Even fragments of casuarina branchlets, when valued as a matter of care through literary art, create micro-openings to a dialogic space where trans-scalar imaginary and arts of noticing are at work.

NOTES

1. This chapter is the culmination of years of reflection on the tree-human relationship on the island of Jinmen. The author is immensely grateful to Professor Iping Liang, whose scholarly zeal and engagement in Critical Plant Studies motivated this research project.

2. Common names include: Australian pine, beefwood, coastal she-oak, and horsetail tree.

3. All translations from Jun-Yao Wu's original Chinese texts are my own.

4. For instance, according to the recorded numbers from Jinmen County Forestry Bureau, 5,619,655 trees were planted in 1966. Among the 5 million new trees one year prior to Wu's birth, casuarinas accounted for a total of 1,523,541. In 1961 and 1971, 2,159,673 and 2,551,678 casuarina trees were planted, respectively. See Statistical Table 11 from *4 Decades of Forest Management in Jinmen*, 41.

5. Scott Slovic, "Liquid Scale: Trans-Scalar Thinking and the Perception of Water," 11.

6. Anna Tsing, "On Nonscalability: The Living World Is Not Amenable to Precision-Nested Scales," 505.

7. Tsing, "On Nonscalability," 506.

8. Yi-Fu Tuan, *Space and Place: The Perspective of Experience*. Minneapolis, 49.

9. The word "curate" derives from the Latin "*curare*," which denotes "take care of."

10. Tuan, "Epilogue: Home as Elsewhere," 227.

11. Jun-Yao Wu, *Three Tree Friends*.

12. Published in 1568, *The Chronicles of the Open Water* was the first book dedicated to historical documentation of Jinmen's agriculture, geographical, and climate features.

13. Shou Hong, *The Chronicles of the Open Water*, 117.

14. In "Militarization and Jinmen (Quemoy) Society, 1949–1992," Michael Szonyi puts the chronological timeline for this period in Jinmen's history in the following order: "(1) ad hoc militarization (1949–1956), (2) formalization and institutionalization (1956–1992), (3) development militarization (1960–1968), (4) combat militarization (1968–1980s), and (5) demilitarization and memory (1990s to present)" (83).

15. "Kinmen Today," https://taiwantoday.tw/news.php?unit=12,29,33,45&%20post=23512.

16. Ibid.

17. Donald Kirk, "Still Waiting for War," https://www.nytimes.com/1976/11/28/archives/ still-waiting-for-war-a-kind-of-tranquillity-has-settled-on-quemoy.html.

18. Slovic, "Language Matters: Environmental Controversy and the Quest for Common Ground," 5.

19. Wu, *Jinmen*, 2.

20. Wu, *I Wish I Was There*, 70.

21. Ibid.

22. Slovic, "Language Matters," 7.

23. Aylan Kurdi's Kurdish name was Alan Kurdi. See https://www.bbc.com/news/world-europe-34141716.

24. Tsing, *The Mushroom at the End of the World: On the Possibility of Life in Capitalist Ruins*, 37.
25. Ibid.
26. Szonyi, *Cold War Island: Quemoy on the Front Line*, 130.
27. See Appendix Table A. 1 in *Cold War Island*, 257.
28. Scott Slovic and Paul Slovic, eds, *Numbers and Nerves: Information, Emotion, and Meaning in a World of Data*, 8, emphasis in original.
29. Ibid.
30. Tuan, *Space and Place*, 162.
31. Ibid., emphasis added.
32. Caitlin DeSilvey, *Curated Decay: Heritage beyond Saving*, 3.
33. Szonyi, *Cold War Island*, 132.
34. Wu, *Jinmen*, 14.
35. Wu, *Jinmen*, 15–16.
36. DeSilvey, 35.
37. Wu, *Maps of Heat*, 226.
38. Tuan, *Space and Place*, 162.
39. Wu, *Maps of Heat*, 194.
40. DeSilvey, 187.

BIBLIOGRAPHY

DeSilvey, Caitlin. *Curated Decay: Heritage beyond Saving*. Minneapolis: University of Minnesota Press, 2017.

Hong, Shou. *The Chronicles of the Open Water* (滄海紀遺). Re-edited by Dao Wu. Taipei: Taiwan Classics Publishing House, 2002.

Hsu, Wei-Min, et al. *4 Decades of Forest Management in Jinmen* (金門林業經營四十年回顧與展望). Jinmen County Forestry Bureau, 1998. "Kinmen Today." *Taiwan Today*, 1 March. 1960, https://taiwantoday.tw/news.php?unit=12,29,33,45&%20post=23512.

Kirk, Donald. "Still Waiting for War." *New York Times*, 28 Nov. 1976, https://www.nytimes.com/1976/11/28/archives/still-waiting-for-war-a-kind-of-tranquillity-has-settled-on-quemoy.html.

Slovic, Scott, and Paul Slovic, eds. *Numbers and Nerves: Information, Emotion, and Meaning in a World of Data*. Corvallis: Oregon State University Press, 2015.

Slovic, Scott. "Liquid Scale: Trans-Scalar Thinking and the Perception of Water." In *Water in Social Imagination: From Technological Optimism to Contemporary Environmentalism*, edited by Jane Costlow, Yrjo Haila, and Arja Rosenholm, 11–27. Leiden: Brill, 2017.

Slovic, Scott. "Language Matters: Environmental Controversy and the Quest for Common Ground." *The Public Land and Resources Law Review* 39 (2018): 1–21.

Szonyi, Michael. *Cold War Island: Quemoy on the Front Line*. Cambridge: Cambridge University Press, 2008.

Szonyi, Michael. “Militarization and Jinmen (Quemoy) Society, 1949–1992.” In *Beyond Suffering: Recounting War in Modern China*, edited by James Flath and Norman Smith, 80–106. Vancouver: UBC Press, 2011.

Tsing, Anna Lowenhaupt. “On Nonscalability: The Living World Is Not Amenable to Precision-Nested Scales.” *Common Knowledge* 18, no. 3 (2012): 505–524.

Tsing, Anna Lowenhaupt. *The Mushroom at the End of the World: On the Possibility of Life in Capitalist Ruins.* Princeton: Princeton University Press, 2015.

Tuan, Yi-Fu. *Space and Place: The Perspective of Experience*. Minneapolis: University of Minnesota Press, 1977.

Tuan, Yi-Fu. “Epilogue: Home as Elsewhere.” In *‘Heimat’: At the Intersection of Memory and Space*, edited by Friederike Eigler and Jens Kugele, 226–238. Berlin: De Gruyter, 2012.

Wu, Jun-Yao. *Jinmen* (金門). Taipei: Erya Publishing House, 2002.

Wu, Jun-Yao. *I Wish I Was There* (如果我在那裡). Taipei: Linking Publishing Co, 2003.

Wu, Jun-Yao. *Three Tree Friends* (三位樹朋友). Jinmen County Cultural Affairs Bureau, 2010.

Wu, Jun-Yao. *Maps of Heat* (熱地圖). Taipei: Chiu Ko Publishing, 2014.

Chapter 7

Mangrove Taiwan

A Birdman's View of an Island Environmental Identity

Rose Hsiu-li Juan

This essay highlights the significant role of mangroves in Taiwan as a lens to explore the island" consciousness and identity. Along the coastal frontiers, mangroves thrive in wetlands, lagoons, and sandbars where rivers meet the sea. Abundantly growing, these mangroves form a distinct boundary, delineating Taiwan as an island. Considering their geo-ecological and geopolitical importance, I contend that mangroves play a crucial role in shaping Taiwan's island identity.

The essay commences by delving into the world of mangroves, shedding light on their species, distribution, recognition of their ecological contributions, and the challenges associated with their destruction and restoration, with a particular emphasis on the context of Taiwan. The environmental, political, and mangrove preservation movements in Taiwan are intricately intertwined and aligned with global progressive trends. Furthermore, this essay portrays the "mangroves Taiwan" through the lens of "Birdman" Liu Ka-shiang. As a birdwatcher, writer, and environmental activist, Liu provides unique insights from both the birdwatcher's perspective and the birds' viewpoints, skillfully presented in his nature and creative writings. In conclusion, while Taiwan is not the primary habitat for mangroves as seen in tropical islands, the unique ecology of mangroves and their intimate connection with seawaters play a crucial role in fostering a heightened recognition of Taiwan's identity as an island. This emerging consciousness stands in contrast to the continental ideology brought by the Han Chinese settlers from mainland China.

AN OVERVIEW OF THE MANGROVE

Mangroves refer to tropical intertidal forests and the tree species that inhabit these unique ecosystems. These ecosystems are primarily situated in tropical and subtropical regions globally, with a notable presence in Southeast Asia, extending northward as far as Kyushu Island, Japan. According to the latest angiosperm classification system (APG IV) and the comprehensive report by Saenger et al. in 2019, the world is home to a total of 81 mangrove plant species distributed among 17 mangrove families. These families exhibit diverse characteristics, with some species featuring areal shoots and buttresses (Sheue and Yong 2020, book description).

However, the distribution of mangroves around the world is highly uneven, marked by a distinct east-west divide. The Indian Ocean-Western Pacific region boasts a remarkable 69 species of mangroves, which is significantly more diverse compared to the 13 species found in the Atlantic Ocean-Eastern Pacific region. In the case of Taiwan, China, and Japan, these regions exhibit relatively fewer mangrove species, and there is a lack of comprehensive reports and data on this topic (Sheue and Yong 2020, book description). In acknowledgment of the pivotal role mangroves play in ecology, UNESCO has designated July 26 as World Mangrove Day. Mangrove forests emerge as highly effective carbon storage systems, playing a pivotal role in mitigating climate change, countering the effects of extreme weather, and addressing sea-level rise. Beyond their environmental impact, mangroves contribute to a rich and diverse ecosystem, offer shoreline stabilization, diminish the impact of natural disasters, support local livelihoods, and provide vital habitats for aquatic life. In alignment with global conservation efforts, Taiwan actively participates in the campaign for mangrove restoration.

MANGROVES IN TAIWAN: DESTRUCTION AND RESTORATION

The tidal wetlands, home to mangrove forests, were often dismissed as swampy wastelands. Without adequate preventive measures, these areas became frequent targets of illegal occupation and dumping grounds for garbage. In a presentation titled "Mangrove and Megacity: War and the Building of Saigon's Urban Archipelago," David Biggs highlighted the mangrove devastation during the Cold War:

> The mid-twentieth century marks a global takeoff in climate-impacting activity, such as urban growth and deforestation that are part of the Great Acceleration. In Southeast Asia, colonial ports in the 1950s grew into vibrant capitals that

> became platforms for new leaders and sometimes battlegrounds in the Cold War. Much of this urban growth came at the expense of coastal ecosystems, especially mangroves, as naval engineers built ports and migrants established settlements along canals and riverbanks choked with waste. (Biggs 2022, lecture description)

Biggs, an environmental historian, is presently conducting research on mangroves and urban-coastal interfaces, not only in Vietnam but also across monsoon Asia. Through these endeavors, he aims to produce a book with the same title as the lecture.

Similar tales of mangrove loss unfolded in Taiwan. The inner sea of Kaohsiung Harbor once hosted the largest expanse of mangrove forests in Taiwan. However, during the Harbor's phase II expansion in 1969, all these mangrove forests were completely eradicated. The story repeats itself near metropolitan Taipei, where the development of high-rise structures along the downstream Tamsui River led to the indiscriminate felling of numerous mangroves to make way for construction purposes. These are common scenarios of mangrove degradation in various regions worldwide.

THE ENVIRONMENTAL MOVEMENT IN TAIWAN: TAKING YI-KUNG MA (馬以工) FOR EXAMPLE

In Taiwan, the environmental movement of the 1960s bore witness to protests against the destruction of mangroves and calls for mangrove conservation, aligning with the growing global concerns for the environment, particularly regarding mangroves. At the forefront of this movement was Yi-kung Ma, a journalist and trailblazer in cultural heritage preservation and environmental protection. Her co-authored book with Han Han (韓韓), titled *We Have Only One Earth*, played a pivotal role in catalyzing environmental activism during that era. Yi-kung Ma's advocacy extended to the preservation of the Tamsui River Mangrove Wetland, protection of the Northeast coast, safeguarding Kenting migrant birds, and the establishment of national parks. Through the influential mass media, her literary journalism served to awaken a generation of Taiwanese people to the importance of environmental protection.

After decades of promotion, mangroves in Taiwan have transitioned from being geographically peripheral to occupying a prominent place in cultural discourse. The government has taken significant strides by initiating conservation and preservation projects, leading to the establishment of nature reserves for mangroves. These reserves have now evolved into popular environmental recreational areas suitable for people of all ages, providing on-site environmental education for both the young and old alike. Expanding its commitment beyond national borders, the Taiwanese government leverages its conservation experience by exporting and sponsoring mangrove restoration projects in countries such as Mozambique and Tanzania. Scholars, backed by government funding, conduct research projects on mangrove studies in West Pacific island nations. Concurrently, private enterprises contribute to the cause by donating money from each purchase to plant mangroves in Tuvaru, an island nation grappling with the effects of global warming. Remarkably, mangroves have not only become a crucial part of Taiwan's environmental identity but have also emerged as cultural capital in the nation's diplomatic relations, fostering island-to-island connections and establishing bonds with countries rich in mangrove ecosystems.

TAMSUI/DANSHUEI RIVER MANGROVE NATURE RESERVE: AN ECOLOGICAL MODEL

Among the diverse mangrove reserves, the Tamsui River Mangrove Nature Reserve in North Taiwan stands out for its superb location and exceptional beauty. Positioned at the confluence of the Tamsui and Keelung Rivers, this Reserve features a Pure Forest predominantly inhabited by the homogenous mangrove species Kandelia Candel/Kandelia obovata, affectionately nicknamed the "water pen" (水筆仔 shuei-bi-zhai in Mandarin Chinese) owing to the tree's candle-like viviparous seedlings. These seedlings, turning vertical, send their roots downward to anchor in the mud. The enchanting image of these baby plants, born directly from the mother trees, sparks literary imagination and symbolizes creativity. Furthermore, water pens, carried by sea currents, find their way to various locations, symbolizing world travelers unrestricted by boundaries.

The Reserve represents a typical river-mouth wetland where fresh and salt waters meet. Nestled within the marshland is the thriving mangrove ecosystem, a haven for diverse beach-dwelling creatures such as crabs, fish, shrimp, and various water birds—both resident and migratory. Among the inhabitants are favorites among children, including mudskippers and fiddler crabs. Adjacent to the Reserve, the Mangrove Ecocenter adds to the experience by offering relevant exhibitions. It serves as a gentle reminder to visitors that Taiwan, surrounded by seas, is a subtropical island.

Positioned approximately 5 kilometers southward is Guang-du Nature Park (關渡自然公園), complementing the Tamsui River Mangrove Nature Reserve as a guide to Taiwan's ecological relationship with the ocean, specifically through its distinctive mangrove environment. Both the Reserve and the Park attained official designation simultaneously in 1986, marking their recognized ecological importance. Conveniently accessible via the metro (MRT) and in close proximity to Taipei City, these sites have become magnets for numerous visitors seeking mangrove education and ecological recreation. Renowned for their avian diversity, especially during winter when thousands of migratory birds consider the island a sanctuary, both locations are popular hubs for bird-watching enthusiasts.

TAIJIANG NATIONAL PARK: CULTURAL AND HISTORICAL COMMONS

In Southern Taiwan, Taijiang National Park (臺江國家公園) was established in 2009 with multifaceted objectives. In contrast to the Tamsui Reserve's 76.4 hectares of Pure Forest, Taijiang National Park, which is 500 times larger at 39,310 hectares, boasts significantly greater biodiversity. The park aims to conserve not only the wetland's ecological richness but also the history of settlement and the local salt and fishing industries. Inclusively designed, the park focuses on both ecological and cultural landscapes, encapsulating the distinctive culture and history of the region. Scattered throughout the park are tidal flats and sandbanks, lagoons and shoals, salt pans, and fish farms adorned with expansive patches of mangrove forests. The rivers meander gently through the Chiayi-Tainan Plain, eventually reaching the sea. This slow flow accumulates a substantial amount of sediment, giving rise to alluvial fans, sandbars, and barrier islands within the breaker zone.

The park is home to four existing species of mangroves, while two other species are already extinct. A notable attraction within the park is the boat tour of the Sicao Mangrove Tunnel, a segment of historical canals that showcases forgotten waterways. The Tunnel is located in the Sicao Wetland at the southern end of the Taijiang Inland Sea. Spanning 547 hectares, the Wetland was formed, along with two other lagoon districts, Kunshen and Cigu, due to massive floods resulting from a storm in 1823 that dramatically altered the course of the Zengwen River. Historically known as the Blind Serpent, the Zengwen River changed its course drastically four times. The considerable sediment carried downstream during these changes deposited and created new lands in the Taijiang Inland Sea, shaping the landscape we see today. The once two inland seas disappeared, and the bustling downtown Tainan was once waterfront property. Wang Chia-hsiang's historical novel, *Dao-feng Inland Sea* (倒風內海 Daofeng neihai), published in 2005, vividly depicts the Indigenous Siraya people navigating the labyrinth of swampy mangrove forests in the seventeenth century. The once awe-inspiring and fearsome mangrove forests have long faded from the memory of the Taiwanese people.

LIU KA-SHIANG: A BIRDMAN'S PERSPECTIVE ON MANGROVE TAIWAN

While the mangrove forests in Wang's *Dao-feng Inland Sea* are part of a historical fantasy, we gain a "birdman's" view of today's mangroves through Liu Ka-shiang. Liu, nicknamed the Birdman, earned this moniker due to his lifelong passion and dedication to birdwatching and ecological preservation. Mangroves take center stage in the coastal wetlands where migrant water birds flock. In *The Vanishing Subtropics* (1986), Liu remarks, "If the cattle egrets stop coming, it means that the mangroves are gone. If the mangroves disappear, the entire Tamsui River becomes a dead river" (Liu 1986, 218).

Liu embarked on his creative journey when he joined the Hua-gang Poetry Society at Chinese Culture University. Among the young poets, he stood out as the most reserved. While his peers engaged in lively discussions, Liu often remained deep in thought, occasionally sighing and often standing alone or pacing with a melancholic expression. Affectionately called "Birdie" (鳥仔tsiáu-á), he seemed to "perch" in the gathering rather than socializing with his associates. It wasn't until his compulsory military service in the navy that Liu, captivated by black herons in Penghu waters, began his journey into bird watching and bird studies and immersed himself in nature and travel writing. Over the next couple of years, true to his initial nickname "Birdie," Liu evolved from a reserved young man into an influential poet, essayist, and ecologist, earning the nickname "Birdman" (Lin 2021, Facebook 2021). With this, he secured a unique niche in the Taiwanese literary world. He explored numerous corners of Taiwan, documenting firsthand experiences and observations made with his own eyes and feet. Refusing to adhere to a single writing mode and rarely repeating himself in content, as Wu Ming-yi aptly describes, "He does not confine himself to a particular writing style and seldom repeats himself in content" (Wu 2011, 128). Liu writes poetry, essays, novels, reports, bird guides, natural history, historical trail guides, travelogues, and picture books. His diverse subjects range from geography and cultural history to insect and vegetable observation. All of his works are innovative and infused with a sense of mission to propel Taiwan toward a better place for both humans and nonhumans. His numerous works have become influential across various fields.

Vulnerable and mostly neglected, wetlands are subjected to large-scale developments and constructions, commonly allocated for building incinerators, landfills, and thermal power plants. Overwhelmed by the environmental predicament, Liu rose up to question developmentalism and spoke for the voiceless wetlands. Through years of close observation, Liu learned how birds are dependent on wetlands. He transformed from a solitary bird watcher who turned his back on society to a social activist actively involved in ecological protection and joined the burgeoning environmental movement. What begins as a bird lover's concern for the survival needs of his dearly beloved turns into a lifelong commitment to championing environmental causes. Between 1982 and 1984, Liu documented the migration of waterfowl from the mouth to the upper reaches of the Tamsui River, simultaneously witnessing environmental deterioration along the northern coast. His emotionally charged prose collection, *Following the Birds to the End of the World*, expresses deep resentment and remorse for the vanishing wetlands.

The Guandu Wetland was a central focus of Liu's investigation along the Tamsui River. Situated at the confluence of the Tamsui River and the Keelung River, the Guandu Wetland has evolved into the largest waterfowl and water bird relay station in northern Taiwan. Unfortunately, during that period, a lack of ecological awareness among both the government and the public resulted in the wetland suffering from reckless dumping of waste soil, indiscriminate logging of coastal forests, and the illegal construction of pigsties and duck farms. These activities posed a significant threat to the habitat for migratory birds. The wetland was in grave danger, rapidly disappearing due to human-induced destruction. Creative essays such as "The Relay Station for Migratory Birds," "Guandu Wetland Over the Past Century," and "Two Stone Poles Rescuing a Piece of Wetland" conveyed Liu's sense of urgency, concern, and a call for action.

Mangroves play a crucial role in stabilizing ecosystems and preserving biodiversity. They are integral to wetland ecosystems, providing essential environments for the migration and nesting of numerous bird species, as well as serving as a habitat for the growth and development of various marine organisms. The mangrove ecosystem is vital for the overall stability of the ecosystem and the maintenance of biodiversity.

Likewise, mangroves play a fundamental role in Liu's self-education about Taiwan's island environment and become a hallmark of his guided walking tours of wetlands. Those who followed him into the mangrove wetlands during his guided tours were deeply impressed by Liu's enthusiasm at the Tamsui River mouth. He shared stories, among others, about the ups and downs of mangroves since he first fell in love with them.

While mangroves are an indispensable part of the wetlands where Liu explores for birdwatching, they don't occupy as conspicuous a position in his environmental writings as they do in real-life events. Wetlands, often located in remote areas far from human habitation, are nestled into the landscape and not easily noticeable. They are as "natural" in the coastal scenery as all other groves where birds pass by, rest, feed, and propagate. Concerned about the overall bird ecology, Liu does not often distinguish mangroves from the whole. In the Birdman's eyes, mangroves do not stand independently from the coastal ecosystem. The ubiquity of mangroves is paradoxically explained by their anonymity.

The following offers an oblique view of mangroves, whose occasional presence surfaces in the text to deter forgetfulness and attest to their existential being in the wetland ecosystem. Exemplifications from Liu's prose collection *Following the Birds to the End of the World* (1985) vividly and poetically document water birds and their behavioral patterns. He depicts golden plovers (金斑鴴), greenshanks (青足鷸), sandpipers (濱鷸), etc. Each species and each flock have its own pace. Between the ebbs and floods, sometimes they stood by or fed at the sandbars of mangrove forests; other times, they flew over to different trees (Liu 1985, 34–35, 43, 45, 54).

Liu's intimate and personal study in the field ignited his interest in nineteenth-century travelers from abroad who, much like himself, observed Formosa with keen eyes and bodily involvement. In the book *Across Formosa: Foreigners' Travel in Taiwan (1860–1880)* (1989), he imaginatively resurrected six historical figures as travelers exploring various parts of Taiwan in traditional ways, either on foot or on vehicles carried by man. Mangroves made an appearance in the chapter about M. Beazeley's geographic investigation as a member of the Royal Geographical Society of London in the southern tip of Taiwan.

In Beazeley's travel journal, Liu points out several notable natural geographic features, including the mangrove forests in Cihou (旗后) Peninsula and Checheng (車城), as well as substantial populations of macaque monkeys in the area, and the peculiar tropical rainforest and short grassy slopes in Kenting (墾丁) (Liu 1989, 121). As he departed from Takao (打狗) and headed south via the Cihou Peninsula, Beazeley recounted: "As we journeyed along the coast and traversed the mangrove forests, the seawater reached up to the knees of the bearers. We passed through several lush fields of indigoes. In other areas of the sandbars, there were dense clusters of screw pines, mango trees, bamboo, and Sago Palms, among others. Attempts to force through there was impossible" (Liu 1989, 122). Eventually, they "took the right path, crossed over sand dunes, and arrived at the beach, escaping the mangrove swamp" (Liu 1989, 123). In a parenthetic note, Liu mentions that this extensive mangrove forest has since vanished (Liu 1989, 123).

In the prose collection *A Flying Squirrel's View* (2004), Liu recounts a journey to the Tamsui estuary:

> Embarking on our voyage, we left the bustling city and the routines of daily life behind. The barge guided us through a mosaic of murky marshlands, across withered brown grasslands, into the vibrant green embrace of the mangrove forest, and finally, we reached the expansive golden sand bars. In spring, beneath

> the warm sunshine and accompanied by flocks of water birds at the estuary, a steady and warm palette emerged, lingering and awaiting our arrival. (144)

Mangroves gleamed in green within this colorful and sunlit world. Liu writes about his surroundings with lyricism and tenderness, mirroring the way he anthropomorphizes animal characters in the book, such as a sea turtle, the piglet Cimino, a sperm whale, and, of course, the squirrel mentioned in the book's title. Much like Liu, these animals are "terrans" (earth dwellers) with a profound attachment to the world they inhabit.

Mangroves find a due place in Liu's fictional works besides playing a significant role in his personal narratives and historical accounts. In the novel *The Wind Bird Pinocchia*, the Tamsui River Wetland unfolds through the perspective of migrant birds, particularly the eastern ring-necked plovers, affectionately nicknamed the "wind birds." In search of Black Shape, the missing hero plover, Pinocchia and Limp Foot fly to a mangrove swamp at the bustling confluence of two rivers. While there are more species of water birds compared to the large sandy area, they find only a few plovers there because the food available in the more open space does not suit their taste (Liu 2007, 65). Mangrove swamps appear as a common feature in the estuary wetlands. In another scene, Pinocchia and Manan, pressing close to the water surface, swiftly trace their way upstream. The scenery on both sides of this stream presents a landscape distinct from the northern river: "There were no lush mangrove forests, nor desolate swamps; instead, there were alternating bamboo clusters, sandbars, reedy banks, and coast hibiscus woods" (Liu 2007, 98). The absence of mangroves becomes notably apparent here.

Among his various works, animal fiction is a noteworthy category. From 1991 to 2008, Liu published four animal fiction works, including *The Wind Bird Pinocchia* (1991), *Hollyann Momo the Humpback Whale* (1993), *Hill of Stray Dogs* (2007), and *Albatross Forever* (2008). Liu employs anthropomorphism to depict various birds and animals and fable-like approaches to enable direct interaction and dialogue between humans and wildlife. Liu knows them well enough and identifies with them. Moreover, realism fails to convey something deeper in his mind aroused by the untamed wildness and vitality of nature out of human control.

In the widely referenced short poem of eight lines, "Exile of the Mangrove Swamp," Liu vividly captures his worries regarding environmental degradation (Liu 2004, 269. Trans. Kaldis, Chinese original 2001). Notably, without explicitly mentioning "mangrove" or its distinctive pen-like seedlings within the poem itself, Liu employs a creative simile: "A cluster of finely-woven green sprouts/ Like the crowding of tiny fishes" to depict the experience of exile. The poem portrays the plight of the baby water pens—Taiwan's dominant mangrove species, renowned for their viviparous nature—likened to live animals. The fish-like mangrove sprouts, floating "against the current" amid "vanishing patches of waves and mounds of garbage," unable to find secure ground for rooting, effectively illustrate the poem's central theme of an imperiled environment. This evocative portrayal not only conveys Liu's deep care and concern for wetlands and coastal ecology but also encapsulates his broader apprehensions about Taiwan's environmental condition.

In Liu's coastal Taiwan, mangroves flourish, commonly found in wetlands along the borders of fresh and saltwater. These mangroves serve as habitats and feeding grounds for water birds, mainly migratory species. Liu's profound sense of Taiwan as an island, with its unique beauty and sorrow, stems from his passion for birdwatching and his keen attentiveness to the environmental nuances associated with the fauna and flora of coastal wetlands.

In an interview with Tsung-Chieh Huang, Liu shared details of his transformative journey. His initial inspiration derived from Rachel Carson's works, including *Under the Sea Wind* (1941) and *The Sea Around Us* (1952). Birdwatching marked the inception of Liu's connection with the natural environment, paving the way for his nature writing characterized by close observation and embodied experience in the 1980s. Prompted by Carson's books, Liu delved into metaphysical questions arising from birdwatching. Recognizing common ground with Carson, he stated, "She stood on a sandy shore and saw mangroves. I stood on a sandy shore and saw mangroves as well. . . . Her shoreline looked much like mine" (Huang 2009, 315). Liu attributed a shift in his perspective to reading Carson's books, noting, "I didn't just go birdwatching; when I walked along the coast, I found myself pondering" (315). Gradually, Liu evolved to develop an island-based nature writing of his own:

> I felt faintly and strongly that I was venturing into a field no past Taiwanese writers had stepped into. . . . While my creative inspiration came from illustrated handbooks of birdwatching from the United States and Japan, I realized that Taiwanese nature writing is distinctly different from that in the United States. (315)

Around 2000, Liu began pondering these questions and came to the realization that his writings were categorically about Taiwan—by Taiwan and of Taiwan—in a way that was unlike anything else in the world. In 1983, Liu authored the fiction *The Wind Bird Pinocchia*, drawing inspiration from his field study at Guandu Wetland and the Tamsui estuary. The simple reason for writing this book at that time was his belief that young people needed native animal stories. During those years, most animal stories were, like fairy tales, imported from abroad with a foreign environmental background. As Liu remarked, "Animal stories have their own boundaries" (318). In response, he crafted tales about stray dogs on the street, whales, albatrosses, and wind birds—subjects taken for granted in Taiwan and experienced uniquely in Taiwan.

Beyond being a pioneer, Liu has continuously committed to writing, consistently pursuing innovations and embracing changes over the years. Liu Ka-shiang is unquestionably a key figure for anyone looking to comprehend the ongoing evolution of nature writing in Taiwan.

AN ALTERNATIVE IMAGINARY OF TAIWAN AS A MANGROVE ISLAND

At the convergence of rivers and seas, where avian congregations gracefully gather, mangroves emerge as crucial architects, shaping estuaries and coastal wetlands. These mangrove wetlands serve as gracious waystations for migratory bird visitors. Liu Ka-shiang, renowned as the Birdman, astutely observes the rhythmic ebb and flow of these feathered guests who seasonally grace Taiwan with their presence. His vigilant watch encompasses the diverse array of animals with the mangroves steadfastly anchoring the backdrop of this ecological narrative. In Liu's portrayal of Taiwan, these mangroves become integral characters along with creatures both large and small. Through his island-based lens, Liu weaves a narrative that offers an insightful understanding of Taiwan's ecological tapestry, where mangroves play a silent yet indispensable role.

While Taiwan's lush forests receive expansive attention, mangrove forests, though recognized as environmental symbols and hailed as paradises for birds and tidal animals, are often relegated to limited literary and socio-cultural acknowledgment. There exists a certain reserve among Taiwanese people when approaching mangroves, with many halting at the edges of the swampy forests. This stands in stark contrast to the deep affection people hold for mountain forests.

Whether experienced firsthand or through Liu's evocative words, we are encouraged to take a profound leap of imagination, envisioning Taiwan as an island-based environment. Liu, in his exploration, discovers a strong sense of Taiwanese identity through encounters with lives that animate Taiwan as an island, with mangroves standing as one of the cornerstones. Within Liu's prominent nature writings and fiction work, mangroves emerge as a testament to the island's distinct ecological character, fostering a deep affinity with tropical and subtropical islands where these unique ecosystems thrive.

Overall, the intricate interplay of environmental, historical, and literary dimensions surrounding mangroves serves as a palpable force in shaping Taiwan's distinctive island identity. Arising from the marshy, muddy, and salty tidal zones is a grounded and place-based environmental consciousness that not only defines Taiwan's island character but also draws parallels with mangrove-clad islands across tropical and subtropical regions worldwide.

REFERENCES

Biggs, David. 2022. "Mangrove and Megacity: War and the Building of Saigon's Urban Archipelago." Lecture. Rachel Carson Center, Deutsches Museum, München. July 25, 2022. Lecture description: https://www.deutsches-museum.de/forschung/veranstaltung/mangrove-and-megacity.

Carson, Rachel. 1952. *The Sea Around Us.* Oxford: Oxford University Press.

Carson, Rachel. 1941. *Under the Sea-wind.* New York: Simon & Schuster.

Huang, Tsung-Chieh. 2009. "The Concept of Nature in Liu Ka-shiang's Animal Fictions" (劉克襄動物小說中的自然觀). *Dong Hwa Journal of Chinese Studies*, 285–324. Accessed August 18, 2023. http://faculty.ndhu.edu.tw/~dhjcs/text/textvol10/vol10.9.pdf.

Lin, Chi-Yang (aka Xiang Yang). 2021. "Liu Ka-shiang the 'Birdman,' an All-round Author." (全能的「鳥人」作家劉克襄). *Facebook.* Last Edited March 14, 2021. Accessed August 15, 2023. https://www.facebook.com/notes/356517778755881/. Original publication: Xiang, Yang. 2018. "'Author's Letters': Liu Ka-shiang the 'Birdman,' an All-round Author." (〈作家的批信〉：全能的「鳥人」作家劉克襄). *WenHsun* 395: 102–107.

Liu, Ka-shiang. 1989. *Across Formosa: Foreigners' Travels in Taiwan (1860–1880)* (橫越福爾摩沙:外國人在台灣的旅 行 (1860–1880)). Taipei: Zi Li Wan Bao.

Liu, Ka-shiang. 2008. *Albatross Forever* (永遠的信天翁). Taipei: Yuan-Liou Publishing.

Liu, Ka-shiang. 2004. "Exile of the Mangrove Swamp" (紅樹林的流亡). Translated by Nick Kaladis. *ISLE: Interdisciplinary Studies in Literature and Environment* 11, no.2 (Summer): 269. Accessed February 15, 2021. https://doi.org/10.1093/isle/11.2.269. Chinese original in *The Most Beautiful Time* (最美麗的時候). Taipei: Titan Publishing Co., Ltd., 2001. 51.

Liu, Ka-shiang. 1985. *Following the Birds to the End of the World* (隨鳥走天涯). Taipei: Hongfan Bookstore.

Liu, Ka-shiang. 2007. *Hill of Stray Dogs* (野狗之丘). Taipei: Yuan-Liou Publishing.

Liu, Ka-shiang. 1993. *Hollyann Momo the Humpback Whale* (座頭鯨赫連麼麼). Taipei: Yuan-Liou Publishing.

Liu, Ka-shiang. 2004. First Published 1984. *A Flying Squirrel's View* (小鼯鼠的看法). Taichung: Morning Star Publishing.

Liu, Ka-shiang. 1986. *The Vanishing Subtropics* (消失中的亞熱帶). Taichung: Morning Star Publishing.

Liu, Ka-shiang. 2007. First Published 1991. *The Wind Bird Pinocchia* (風鳥皮諾查). Taipei: Yuan-Liou Publishing.

Ma, Yi-kung, and Han Han. 1983. *We Only Have One Earth* (我們只有一個地球). Taipei: Chiu Ko Publishing.

National Museum of Taiwan Literature. 2017. *The Liu Ka-shiang Archive*. https://enlkshiang.nmtl.gov.tw/

Sheue, Chiou-Rong, and Jean Wan Hong Yong. 2020. *Exploring the Sea Forest: An Introduction to the Indo-West Pacific Mangroves* (探訪海上森林:印度-西太平洋紅樹林簡介). Taichung: National Chung Hsing University Press. Book description in NCHU library catalog. Accessed June 16, 2023. https://nchu.primo.exlibrisgroup.com/discovery/fulldisplay?docid=alma990072721140107976&context=L&vid=886NCHU_INST:886NCHU_INST&lang=zh-tw&search_scope=MyInst_and_CI&adaptor=Local%20Search%20Engine&tab=Everything&query=any,contains,%E7%B4%85%E6%A8%B9%E6%9E%97

Wang, Chia-hsiang. 1997. *Dao-feng Inland Sea* (倒風內海). Taipei: Taiwan Interminds Publishing.

Wu, Ming-yi. 2011. First Published 2003. *Essays by Taiwanese Nature Writers 1980–2002: Liberating Nature through Writing, BOOK 2* (臺灣自然書寫的作家論1980~2002：以書寫解放自然BOOK 2). New Taipei: Summer Festival Press.

Chapter 8

Politics of Femininity, Politics of Plants

The Roof Garden in Zhu Tianwen's "Fin de Siècle Splendor"

Pei-Wen Clio Kao

In response to the Anthropocentric tendency of "plant blindness," this chapter aims to shed light on Zhu Tianwen's "Fin De Siècle Splendor" from the lens of Critical Plant Studies (CPS).[1] Former critical approaches have emphasized its lyrical and feminine style, its postmodern mentality, its political allegory, and its materialist indulgences.[2] Among the abundant critical attention that the acclaimed story has received, none of which has touched on the thematic, as well as the aesthetic rendering, of the plant and flora scenes in the story. To fill in the critical gap and make the classic work relevant to the Anthropocene, I borrow ideas from critics in Critical Plant Studies to tease out the desire, agency, and subjectivity of the plant images and their interconnectivity and entwinement with the female protagonist's identity formation, sexual awakening, and gender empowerment.

The narrative of "Fin de Siècle Splendor" centers on Mia, a twenty-five-year-old professional model who has passed the prime of her physical beauty. She is the mistress of an architect, Duan, who is like an avuncular lover to Mia in terms of their age gap. Mia builds an artificial garden on top of her apartment's corrugated roof, which is a common illegitimate structure in Taipei. On the surface, the narrative is anti-chronological and cluttered with bits and pieces of fashion items. Underneath, the story is arranged by time and space specificity: the timeline follows the fashion industry development, and the space is occupied by an array of plants and flowers kept in

Mia's roof garden. The storyline is overwhelmed by the dazzling aesthetics of fashion items as well as catalogs of plants and flowers. It creates a transformative space for Mia and her group.

By probing into the rooftop garden, this chapter is divided into six sections. It begins with "The Platonic Conception of the '*Pharmakon.*'" The second section examines "Mia's Roof Garden and Her Role of the '*Pharmarkeus.*'" The third section analyzes the garden as "Mia's Feminine Kingdom" to be followed by the investigation of "The Human-Plant Intimacy" in the next section. The penultimate section, "Politics of Femininity, Politics of Plants," unpacks the entanglement of gender and vegetal politics. In the conclusion, I explore the vegetal-political meaning of the roof garden in the context of Taiwan-China confrontation.

I start with Jacques Derrida's notion of "*pharmakon*" to untangle the many threads of my argument.[3] According to Derrida, the Greek semantic origin of *pharmakon* denotes either a drug or a poison; a cure of illness or its cause. The undecidability and multivalence of *pharmakon* is also materialized in its word family—*pharmakon* (remedy), *pharmakeus* (sorcerer, magician), and *pharmakos* (scapegoat). In the story of "Fin de Siècle Splendor," Mia builds an "illegal structure" of a "corrugated iron" roof on top of her apartment's "sun terrace," where "the dried flowers are hanging upside down."[4] Like a *pharmakeus* figure, or a witch, the female protagonist collects a rich array of dried plants and flowers in her apartment building, attempting to distill their smell and fragrance and retain their memories. This *pharmakeus* figure is reminiscent of Shennong, or the "Divine Farmer" in Chinese mythology. With a view to tasting plants to make herbal medicine, Shennong is killed by the poisonous herbs during the process. I argue that Mia is a veritable *pharmakeus* or a female Shennong. She provides spiritual guidance for her friends, bringing them laughter and pleasure. On the other hand, the dried plants preserved and the garden created by Mia are lethal and poisonous, bringing about the downfall of the androcentric kingdom and replacing them with a world of the feminine.

Written at the end of the last century, Chu's story of a feminine plant kingdom simultaneously defies the cluster of ideology that is male-centered, west-centered, and human-centered. By sporting Mia's feminine features and displaying her plant collections, Zhu demonstrates how the feminine agency, identity, and desire can be created based on intimate human-plant relations. Moreover, Zhu's intention goes beyond a political allegory of New Taiwan. I argue that Zhu reclaims the power of creation and desire through the politics of femininity, and also through the politics of plants. Rather than merely reversing the hierarchy of man/woman, West/East, human/other-than-human, the politics of femininity and the politics of plants embrace ambiguity, uncertainty, and hybridity—metaphorically an in-between space embodied in

Mia's roof garden, a "semi-outdoor space" shored up by the illegal structures of corrugated irons against the ruins of the fin-de-siècle Taipei City.

THE PLATONIC CONCEPTION OF THE "*PHARMAKON*"

In Jacques Derrida's interpretation, the Pharmacia of Plato's *Phaedrus* "is also a common noun signifying the administration of the *pharmakon*, the drug: the medicine and/or prison."[5] Endowed with innate "ambivalence," *pharmakon* "can be—alternately or simultaneously—beneficent or malficent."[6] In an analogy, the leaves of writing, or the hemlock is compared to *pharmakon* "to push or attract out of the city the one who never wanted to get out, even at the end, to escape the hemlock."[7] We could extract from the indetermination or undecidability of *pharmakon* the words of "'remedy,' 'recipe,' 'poison,' 'drug,' 'phitler.'"[8] Derrida elaborates and recounts the tale of King Thamus and his vassal Theuth. The King is presented the gift of writing, or *pharmakon*, which is "rejected, abandoned, disparaged."[9] This is the origin of Western dominance by speech, by Logos, or by the Name of the Father. On the other hand, the power of writing as opposed to speech is aimed "to do away with the father for achieving emancipation with complacent self-sufficiency."[10] Accordingly, the "desire for writing" or the accepting of the gift of *pharmakon* is imagined as "a desire for orphanhood and patricidal subversion."[11] The figure of the father is the source of "the good": "chief, capital, and good(s)."[12]

Derrida traces the binary oppositions inherent in writing to other ancient mythologies. Thus we get the genealogy of paired concepts such as "speech/writing, life/death, father/son, master/servant, first/second, legitimated son/orphan-bastard, soul/body, inside/outside, good/evil, seriousness/play, day/night, sun/moon," and the list goes on.[13] The second part of each pair serves as a "dangerous supplement," a kind of "substitute capable of doubling for the dint of representing, repeating, and masquerading" and is "also capable of totally supplanting them and appropriating all their attributes."[14] Derrida brought up the "god of writing" and the "god of medicine" together to demonstrate their common traits—"The god of writing is thus also a god of medicine. Of 'medicine': both a science and an occult drug. Of the remedy and poison. The god of writing is the god of *pharmakon*. And it is writing as a *pharmakon* that he presents the king in the *Phaedrus*, with a humility as unsettling as dare."[15]

Derrida further identifies the *pharmakeus* in the images of a "sophist," a "magician," a "sorcerer," and even a "prisoner."[16] And the word *pharmakeus* bears resemblance, or serves as a homonym, to that of *pharmakos*, which in Greek denotes the meaning of a scapegoat. He signifies "the *evil* and the *outside*, the expulsion of the evil, its exclusion out of the body (and out) of the city."[17] The polarized nature of the *pharmarkos* lies in his capability both to "cure" and to "harm." He "represents evil both introjected and projected."[18] The ritual of the scapegoat is "[beneficial] insofar as he cures—and for that, venerated and cared for—harmful insofar as he incarnates the powers of evil—and for that, feared and treated with caution."[19]

MIA'S ROOF GARDEN AND HER ROLE OF THE "*PHARMAKEUS*"

Mia is likened to the figure of a witch, a sorceress, or a "*pharmakeus*," who wields her dual-edged weapon of writing, or the *pharmakon*, to build up and corroborate her kingdom of a garden filled with dried flowers and plants. The garden is on the top of an apartment, a "sun terrace" where "all sorts of dried flowers are hanging upside down."[20] This is the "semi-outdoor space" based on the "structures of corrugated iron," which is the in-between and liminal space of legal/illegal and legitimate/illegitimate occupied by the figure of the *pharmakeus*.[21] Mia's roof garden produces not only a sense of taste—such as the "peppermint herbal tea"—but also a sense of smell and color. "Mia is a fervent believer in the sense of smell" produced by the fragrance of her dried plants and flowers; "Mia is also dependent on the memories evoked by color."[22] As a mistress of her lover, Duan, Mia wanders back and forth the space of the legitimate/illegitimate, the authentic/the artificial, just like her role of the "*pharmakeus*," commanding the garden of herbal plants and flowers. Derrida reminds us that Plato "refers to painting as a *pharmakon*," for in Greek *pharmakon* is equal to paint—"not as a natural color but an artificial taint, a chemical dye that imitates the chromatic scale given in nature."[23]

Mia is like the *pharmakeus* figure, who gives her *pharmakon* in order not only to cure, but also to poison. She takes care of her roof garden and garners the plants to make "jars and bunches of dried flowers and dried reeds in her house."[24] Like a witch or a female Shennong, she makes herbal tea and herbal medicine and "[drinks] hot ginger tea or [puts] lots of powdered cassia into her cappuccino."[25] With her obsession with herbal medicine and her sensitivity to the aesthetics such as bright colors and fragrant smells ("Tahitian flowers and orange shades by Gaugin,") she also serves as the goddess of Nature embraced by her friends.[26] The narrator says "Like the expensive frankiscense she acted as a cohesive force binding her boyfriends together. She was the one to call for a gathering, and everyone came."[27] She led them to Mt. Yangming, where "in the naked, cool vacuum, the sky was suddenly filled with birdsong."[28] In this natural world, "[frankicesnse] floated in the air; birdsong felt like rain to cover them" while those young men are "making love with Nature."[29] Apart from being a goddess, Mia is

> probably a witch. Her flat filled with dried flowers and herbs, like a pharmacy. Duan often has the illusion that he is with a medieval monk. Her bathroom is planted with Chinese orchids, African violets, potted pineapples, peacock coconuts, and all sorts of nameless ferns. On top of all that there are scores of different bath salts, bath oils, soaps, and shower gels in glaring poisonous colors; the room is like a magician's distillery.[30]

Like a "witch," a "medieval monk," or a "magician," Mia's pharmacy gives the bonding and soothing force to embrace her friends and entice her lover Duan. On the other hand, the "poisonous" nature of her pharmacy will in one day undermine the world dominated by men and rebuild a feminine kingdom from the fragments of "her memory of smells and colors."[31]

In my perspective, Mia is also akin to a "*pharmkaos*," a scapegoat figure who bears the projection and introjection of the reader. In most of the critical analysis of the story, Mia is acclaimed for her strength—her feminist independence and her individual charisma. On the other hand, she is also criticized for her flaws—her commercial desire and her insistent worship of commodity culture. Just like a scapegoat figure, Mia receives both the wanted and unwanted parts in the reader's consciousness. Therefore, Mia has become a "*pharmakos*"—turned from a "*pharmakeus*"—by virtue of her concoction of the herbal medicine in the modern-day store of *pharmakon*.

MIA'S FEMININE KINGDOM

Mia not only makes her herbal medicine and does alchemy like a *pharmakeus*, but also claims her feminine charm like a queen to conquer the kingdom once dominated by men. She builds her feminine kingdom on the basis of color, smell, and fabric to replace the male world full of patriarchal systems and structures such as words and institutions. The narrator says, "When she is old and her beauty faded, Mia will be able to support herself with her handicrafts. The abyssal blue tells her that the world men have built with theories and systems will collapse, and that she with her memory of smells and colors will survive and rebuild the world from here."[32] Not only does Mia break through the male-centered epistemology, she also bids farewell to the homoerotic aesthetic in her total embrace of femininity. And the narrator goes on, "How the times have changed! Even Comme de Garçons has made a clear break with the asexual look that made her name and has joined the camp of sheer femininity."[33]

Mia's stance is clear-cut and unequivocal. She discards both stern masculinity and fuzzy homosexuality. She chooses femininity. She wields femininity as a powerful weapon, and wins over men through her feminine charismas: "She fully intends to take advantage of the fact that she is a woman, and the more feminine a woman is, the more she can get out of men."[34] In the fashion industry, femininity is on the rage and is making its comeback: "Faced with the trend toward femininity, Issey Miyake gave up his three-dimensional cutting and shifted his attention to the use of fabrics. Patterns are pressed onto silk and chiffon to create a steel feel quite different from the original texture, giving a hard edge to soft femininity."[35] We could see that the feminine is not sheer effeminate softness; it also contains rock-like strength and perseverance. On the other hand, Mia discards the same-sex homoeroticism, for "a narcissistic woman may be lovable; a narcissistic man is just unmanly."[36] The feminine charms and glamors come from the colorful, fragrant and soft materials, which has made Mia a "material girl" since "she'd worship things and she'd worship money. Youth and beauty were on her side; she worshiped her own beautiful body."[37] Contrary to the phallogocentric dominance of Reason, Mia's feminine kingdom puts the material and the body into the sacred shrine of worship.

In her relationship with Duan, Mia is like a "queen bee," always busy "[making] money" to hoard her material resources.[38] Duan is seduced by her feminine beauty, and she is "like a nymph resurrected."[39] Very much like the encounter between Ulysses and the Siren, the romantically wandering Duan is to be defeated by Mia, the *femme fatale*. On the other hand, Mia and Duan are several generations apart, and "she's young enough to be his daughter."[40] On the surface, this generational gap seems to give Duan the senior and mature force to control the flippant and immature Mia. Nevertheless, with a turn and twist, we are told that in everyday life Mia is behaving more like a "mother," "as if Duan were just a chattering child."[41] This reversal of the gender hierarchy from the father/daughter or master/mistress to that of mother/child is reminiscent of the story of Jane Eyre, an independent and feminine young lady who grows up to assert herself and take care of her former master and once-dominant lover, Mr. Rochester.

THE HUMAN-PLANT INTIMACY

For urban ecologists and ethnographers, "humans and more-than-human interlocutors" share a sense of belonging and form an affective relationship.[42] They emphasize the "relational turn" and "relational ecologies of belonging" when studying human-nature relationships.[43] Following Donna Haraway's exhortation, urban ecologists aim to engage in an "ethical caring of and for 'signature otherness.'"[44] This kind of urban nature forms "rooted cosmopolitanism," which offers "recreational opportunities and improved mental health."[45] We are encouraged to pay attention to the "affective and intimate engagements between humans and more-than-human others" in the various scenarios of the cityscape.[46] Apart from the affective turn or the human and more-than-human relationship, urban ecologists also attempt to redefine the significance of "agency," "as not simply the capacity for individual, willful action, but rather as the capacity for 'response/ability'—which is also the ability to affect and affected by others."[47]

In the urban scenario of Taipei, Mia forms relationships with places and plants, and searches her sense of belonging in this metropolitan, or cosmopolitan, city. In one of her favorite hangouts, the West Gate District, Mia and her girlfriend Baby conceive of the idea of "opening a [flower] shop someday" while sitting at the McDonald's, a "bright and airy [place] like a greenhouse."[48] In fact, Baby later "opened a flower shop" after divorcing her husband.[49] Instead of getting an emotional attachment to the male-centered world, Mia and her friends turn to the plants, the more-than-human agents, to build their feminine family unit based on the sense of intimacy and belonging. Baby also sells food in the flower shop, which denotes the healing and soothing nature of the store as a place of manna. "Baby was sitting on a black rattan stool; from the back her mature figure looked as stable as a huge piece of rock."[50] The feminine figure of Baby is paradoxically associated with a "rock," implying that the gentleness, strength, intimacy, and endurance are all at once incorporated into the world of femininity, plants, and flora. The healing and soothing nature of plants and flowers make the flower shop the best "rendezvous for her girlfriends" to make connections and nourish their beings.[51] The flower shop is ensconcing in "an expensive area, and all the shops along the lane are small boutiques, Mia can smell the shops; sometimes the smell is evoked by colors, sometimes by the décor and the use of space."[52] This is the alternative space shored up by "smells," "colors," and "the feminine." Mia sees it in the image of the "world's ancient civilizations"—"The flower shop with its complex mixture of scents is like a Byzantine tapestry; the aroma of coffee waffles in the air, recalling the ancient age of handicrafts."[53] The plants and flowers are there to entertain, soothe, and heal, with their multifunctional purposes serving as plants, food, and medicine. The narrator observes, "Joey is responsible for the food served in the flower shop: homemade fruit cakes, cheese pie, oatmeal biscuits, and flower-petal puddings."[54]

Baby's flower shop serves as the hub of their sisterly and fraternal bonds, where Mia's group gets together and takes refuge. Both Mia's roof garden and Baby's flower shop are symbolic of the healing force of plants and the intimate interaction between humans and plants. "Sometimes they go to Mia's place to try her new concoctions of herb tea, and Duan is left sitting in a corner listening to music; there is no way he can break into their endless chit-chat talks."[55] The plant and flower place is the feminine kingdom built up and nurtured by Mia. Duan, as a product of patriarchal culture, has no foothold in that kingdom of feminine intimacies. The dried flowers and plants kept by Mia suggest her desire to immerse in the ephemerality of beauty, love, and time to reach an aesthetic of eternity in her plants. Mia's floral and vegetal kingdom is environed by "several bundles of ferns and flowers hanging at the windows, a clay bowl filled with dried yellow rose petals, and a rattan tray containing the dried peel of lemons, oranges, and mandarins."[56] Mia and Duan's relationship won't last for good while the dried flowers and plants in her dwelling shall never perish. They have excelled at the patriarchal institution of marriage in a gesture to embrace eternity, immortality, and permanence.

POLITICS OF FEMININITY, POLITICS OF PLANTS

In the tradition of Western philosophy and metaphysics, the plant is associated with the feminine and the passive; while the animal is the masculine and the active. The logocentric system is structured into binary oppositions with the self dominating the other. In reaction to the anthropocentric and androcentric principle of Western metaphysics, Luce Irigary in her book *I Love to You: Sketch of a Possible Felicity in History* re-imagines the femininity of the plant, which carries the "possibility of restructuring or even creating feminine subjectivity in a way that would make a real difference to women."[57] It demonstrates Irigaray's feminist thought and her vision of the relationship between the two sexes. She does not envision a reverse of the gender hierarchy of man/woman, but instead calls for a reciprocal engagement and mutual recognition that will lead to an alternative. Irigary argues that the nineteenth-century conception of "vegetative soul" or "vegetable genius" takes a departure from the individualization based on binary dichotomy, but instead seeks the individualization with "the promise of life and growth."[58] Accordingly, the feminine subject with its plant growth metaphor is "neither atomistic nor confrontational" by nature.[59] As opposed to the self-sufficient and linear evolution of animals, plants own the transformative force of "metamorphosis and growth" through the practices of grafting, drying, and sampling, as represented in Mia's roof garden.[60] In addition, the non-binary nature of plants is related to the androgynous style of fashion and the narcissistic tendency of Mia's group, composed of pretty gay men and anti-marital women: "The plant moves beyond the opposition of male and female, for both sexes often exists side by side in the same flower."[61]

Irigaray subverts the metaphysical association of women with the plant for "passivity and unconsciousness" and puts forward a new concept of "efflorescence" to express "a blossoming or blooming forth that cannot be enclosed within the traditional boundaries of embodiment and philosophical discourse, and which may have the possibility of transforming it."[62] Moreover, Irigaray redefines the blossoming vegetal imagery of women:

> One might well wonder if women are closer to the vegetable world than to the animal world, as was claimed by certain ancient philosophers, and particularly by female cultures, although, it is true, in different ways. Could it be that in this proximity there lies an accurate explanation of her relation to passivity? Women's receptivity would not be restricted to her relation to man alone but would extend to the natural economy, especially the cosmic one, with which her equilibrium and growth are more closely associated. Her so-called passivity would not then be part of an active/passive pair of opposites but would amount to attentiveness and to *fidelity* rather than *passivity*. A matter, therefore, not of pure receptivity but of a movement of growth that never ultimately estranges itself from corporeal existence in a natural milieu. In which case, becoming is not cut off from life or its placing. It is not extrapolated from the living nor founded in a dead character. It remains attentive to growth: physiological, spiritual, relational.[63]

Reminiscent of the healing and bond-making force of Mia's dried plants and flowers, the "vegetative soul" is characterized by "rootedness, vulnerability, interdependence, and transformative possibility" in contrast to "a stance of aggressiveness and self-preservation" of the animal-like male individualization.[64] Postmodernist philosophers like Jacques Derrida, Gilles Deleuze, and Luce Irigaray emphasize the "political intervention" and "transformative interpretation" of the vegetative soul and vegetable growth.[65]

In addition, Irigaray's conception of the "efflorescence" is traced back to Aristotle's philosophy of the "vegetative soul," which is associated with "reproduction, particularly with the womb and fetation."[66] The notion of efflorescence is "a figure that implies metamorphosis and indefinite individualization," which aims to criticize the "primacy of traditionally 'animalistic' metaphors that emphasize self-enclosedness, self-preservation, and strict identity over time, qualities that among other philosophical constructs, could be predicated of the fiction of the (masculine) subject."[67] The efflorescence and the vegetative soul of the plant indicate a feminine subject that is receptive to the external environment, transformative in its life forms, and capable of healing the wounds—which displays the politics of femininity and politics of plants, grounded on the force of political intervention and transformation.

CONCLUSION

In conclusion, I further argue that Zhu Tienwen's "Fin de Siècle Splendor," based on the politics of femininity and politics of plants, is a metaphor for the cultural hybridity of Taiwan—a place whose culture "is a hybrid of Chinese, Indigenous, and various foreign cultural imprints."[68] The female subjectivity represented by Mia and her group connotes a life form that is anti-binary and anti-hierarchical, which is opposed to the dichotomous "politics of the center (China) versus the marginal (Taiwan), the material versus the spiritual, and woman versus man."[69] In other words, it embraces hybridity, ambiguity, and inbetweeness of the feminine kingdom against the man-made myth and institution.

At the final part of "Fin de Siècle Splendor," Mia has created her own "handmade paper."[70] With her transformative and interventionary force, she invents a feminine way of writing and a feminine kingdom based on "her memory of smells and colors."[71] Although Mia is like Duan's daughter in terms of age, she plays the maternal role "with the look of a mother, as if Duan were just a chattering child."[72] The gender boundary and hierarchy are finally blurred and disrupted. Moreover, the politics of femininity and politics of plants are also manifested in the genre of "romance" to which the story belongs. Zhu not simply reproduces and replicates the thematic focus of the genre and its language of lyricism but also reinvents a new form for the genre, blending the Western modernist techniques with the Chinese sentimental poetry.[73] Such a boundary-crossing approach is best reflected in Mia's invention of the handmade paper at the end of the story. Mia has become an authoress, writing her story with her own approach, distinct from the male authors of the West, the East, China or Taiwan. Like a female Shannong, she rebuilds a feminine kingdom based on her transformative, creative, and reproductive capability, to shore up against the ruins of patriarchal institutions and piece together anew the rubbles of the Anthropocene.

NOTES

1. Catriona Sandilands, "Plants." *The Cambridge Companion to Environmental Humanities* (Cambridge University Press), 156–157.

2. Shu-Chen Chiang, "Rejection of Postmodern Abandon: Zhu Tianwen's *Fin-de-Siècle Splendo*r." *Feminism/Femininity in Chinese Literature* (Rodopi, 2002): 45–66. Chiang challenges the critical tendency to categorize Zhu's work under the umbrella term "postmodern." Instead, she argues that Zhu's conception of the "postmodern Taiwan" is merely a façade to disguise her real intention to bring back the "repressed voices" of people from different classes, genders, and ethnicities

at this particular historical juncture, particularly the "underpriviledged Mainlanders" (46, 63). Sung-sheng Yvonne Chang, "Chu T'ien-wen and Taiwan's Recent Cultural and Literary Trends." *Modern Chinese Literature* 6, no. ½ (Spring/Fall 1992): 61–84. Chang ascribes Zhu's early-day "romantic sentiment" and inheritance of Eileen Chang's feminine and lyrical and classical style to Zhu's sinocentric cultural identity and political stance (64). In Zhu's latter-day works, "Fin de Siècle Splendor" included, she shifts to a mixed style of classical lyricism and popular vulgarism, a pastiche characteristic of postmodernism. The Taipei City represented in the short story also embodies the postmodern condition of city life. David Der-wei Wang, "'Fin-de-Siècle' Splendor: Contemporary Women Writer's Vision of Taiwan." *Modern Chinese Literature* 6, no. ½ (1992): 39–59. Wang starts with "poetics of marginality" to contend how Zhu fulfills her political agenda of speaking for the "marginalized gender subjectivity" and "marginalized political identity" (39, 58). Therefore, the narrative on fashion items and details s to sugarcoat the serious political allegory of contemporary Taiwan's political status and its emerging national identity. C. Laura Lovin, "Interconnectivities and Material Agencies: Consumption, Fashion, and Intimacy in Zhu Tianwen's 'Fin-de-Siècle Splendor." *Intimacies, Critical Consumption and Diverse Economies* (Palgrave Macmillan, 2015), 60–68. By adopting the New Materialist approach, Loving focuses on the commodities and fashion objects scattered in Zhu's fragmentary narrative and argues for the material agency and the inseparability of the bodily materiality and social construction.

3. Jacques Derrida, *Dissemination* (University of Chicago Press, 1981).
4. Zhu Tianwen, "Fin De Siecle Splendor." *The Columbia Anthology of Modern Chinese Literature* (Columbia University Press, 2007), 388.
5. Derrida, *Dissemination*, 70.
6. Derrida, *Dissemination*, 70.
7. Derrida, *Dissemination*, 70.
8. Derrida, *Dissemination*, 71.
9. Derrida, *Dissemination*, 76.
10. Derrida, *Dissemination*, 77.
11. Derrida, *Dissemination*, 77.
12. Derrida, *Dissemination*, 81.
13. Derrida, *Dissemination*, 85.
14. Derrida, *Dissemination*, 90.
15. Derrida, *Dissemination*, 94.
16. Derrida, *Dissemination*, 117.
17. Derrida, *Dissemination*, 130.
18. Derrida, *Dissemination*, 133.
19. Derrida, *Dissemination*, 133.
20. Zhu, "Fin de Siècle Splendor," 388.
21. Zhu, "Fin de Siècle Splendor," 389.
22. Zhu, "Fin de Siècle Splendor," 388.
23. Derrida, *Dissemination*, 129.
24. Zhu, "Fin de Siècle Splendor," 391.
25. Zhu, "Fin de Siècle Splendor," 391.

26. Zhu, "Fin de Siècle Splendor," 391.
27. Zhu, "Fin de Siècle Splendor," 392.
28. Zhu, "Fin de Siècle Splendor," 392.
29. Zhu, "Fin de Siècle Splendor," 392.
30. Zhu, "Fin de Siècle Splendor," 398.
31. Zhu, "Fin de Siècle Splendor," 402.
32. Zhu, "Fin de Siècle Splendor," 402.
33. Zhu, "Fin de Siècle Splendor," 390.
34. Zhu, "Fin de Siècle Splendor," 391.
35. Zhu, "Fin de Siècle Splendor," 401.
36. Zhu, "Fin de Siècle Splendor," 394.
37. Zhu, "Fin de Siècle Splendor," 393–394.
38. Zhu, "Fin de Siècle Splendor," 395.
39. Zhu, "Fin de Siècle Splendor," 395.
40. Zhu, "Fin de Siècle Splendor," 401.
41. Zhu, "Fin de Siècle Splendor," 401.
42. Melissa R. Poe and Joyce LeCompte, "Urban Foraging and the Relational Ecologies of Belonging." *Social and Cultural Geography* 15, no. 8 (2014): 901.
43. Poe and LeCompte, 902.
44. Poe and LeCompte, 903.
45. Poe and LeCompte, 904.
46. Poe and LeCompte, 904.
47. Poe and LeCompte, 905.
48. Zhu, "Fin de Siècle Splendor," 395.
49. Zhu, "Fin de Siècle Splendor," 396.
50. Zhu, "Fin de Siècle Splendor," 396.
51. Zhu, "Fin de Siècle Splendor," 396.
52. Zhu, "Fin de Siècle Splendor," 396.
53. Zhu, "Fin de Siècle Splendor," 396.
54. Zhu, "Fin de Siècle Splendor," 396–397.
55. Zhu, "Fin de Siècle Splendor," 398.
56. Zhu, "Fin de Siècle Splendor," 399.
57. Elaine P. Miller, *Vegetative Soul: From Philosophy of Nature to Subjectivity in the Feminine* (State University of New York Press, 2002), 4.
58. Miller, *Vegetative Soul*, 4, 5.
59. Miller, *Vegetative Soul*, 5.
60. Miller, *Vegetative Soul*, 8.
61. Miller, *Vegetative Soul*, 11.
62. Miller, *Vegetative Soul*, 16.
63. Miller, *Vegetative Soul*, 190. Luce Irigaray, *I Love to You: Sketch of a Possible Felicity in History* (Routledge, 1996), 38–39.
64. Miller, *Vegetative Soul*, 18.
65. Miller, *Vegetative Soul*, 183. Derrida, *Dissemination*. Gilles Deleuze and Felix Guattari, *A Thousand Plateaus* (Minnesota University Press, 1987), 3–25. These contemporary philosophers imagine the reading process as that of plant growth, which

will sprawl and bloom without a fixed and inherent pattern. Derrida's idea of dissemination; Deleuze and Guattari's conceptual imagery of the rhizome, and Irigaray's metaphor of efflorescence all capture the multivalent and polymorphous nature of the writing and reading of texts.

66. Miller, *Vegetative Soul*, 187.

67. Miller, *Vegetative Soul*, 188, 189.

68. Lingchei Letty Chen, "Writing Taiwan's Fin-de-Siècle Splendor: Zhu Tianwen and Zhu Tianxin." *The Columbia Companion to Modern Chinese Literature* (Columbia University Press, 2016), 364.

69. Chen, "Writing Taiwan's Fin-de-Siècle Splendor: Zhu Tianwen and Zhu Tianxin," 365.

70. Zhu, "Fin de Siècle Splendor," 401.

71. Zhu, "Fin de Siècle Splendor," 402.

72. Zhu, "Fin de Siècle Splendor," 401.

73. Sung-sheng Yvonne Chang, "Yuan Qiongquiong and the Rage for Eileen Zhang among Taiwan's 'Feminine' Writers." *Modern Chinese Literature* 4, no. ½ (1988): 201–223.

BIBLIOGRAPHY

Chang, Sung-sheng Yvonne. "Yuan Qiongqiong and the Rage for Eileen Zhang among Taiwan's 'Feminine' Writers." *Modern Chinese Literature* 4, no. ½ (1988): 201–223.

Chang, Sung-sheng Yvonne. "Chu T'ien-wen and Taiwan's Recent Cultural and Literary Trends." *Modern Chinese Literature* 6, no. ½ (1992): 61–84.

Chen, Lingchei Letty. "Writing Taiwan's 'Fin-De-Siècle Splendor': Zhu Tianwen and Zhu Tianxin." In *The Columbia Companion to Modern Chinese Literature*, edited by Kirk A. Denton, 363–370. Columbia University Press, 2016.

Chiang, Shu-Chen. "Rejection of Postmodern Abandon: Zhu Tianwen's *Fin-De-Siècle Splendor*." In *Feminism/Femininity in Chinese Literature*, edited by Peng-hsiang Chen and Whitney Crothers Dilley, 45–66. Rodopi, 2002.

Deleuze, Gilles, and Félix Guattari, *A Thousand Plateaus: Capitalism and Schizophrenia*. Translated by Brian Massumi. Minneapolis: Minnesota University Press, 1987.

Derrida, Jacques. *Dissemination*. Translated by Barbara Johnson. University of Chicago Press, 1981.

Irigaray, Luce. *I Love to You: Sketch of a Possible Felicity in History*. Translated by Alison Martin. Routledge, 1996.

Lovin, C. Laura. "Interconnectivities and Material Agencies: Consumption, Fashion, and Intimacy in Zhu Tianwen's 'Fin-de-Siècle Splendor.'" In *Intimacies, Critical Consumption and Diverse Economies*, edited by Emma Casey and Yvette Taylor, 60–83. Palgrave Macmillan, 2015.

Miller, Elaine P. *The Vegetative Soul: From Philosophy of Nature to Subjectivity in the Feminine*. State University of New York Press, 2002.

Poe, Melissa R. and LeCompte, Joyce. “Urban Foraging and the Relational Ecologies of Belonging.” *Social and Cultural Geography* 15, no. 8 (2014): 901–919.

Sandilands, Catriona. “Plants.” In *The Cambridge Companion to Environmental Humanities*, edited by Jeffrey Jerome Cohen and Stephanie Foote, 156–176. Cambridge University Press, 2021.

Wang, David Der-wei. “‘Fin-de-Siècle’ Splendor’: Contemporary Women Writers’ Vision of Taiwan.” *Modern Chinese Literature* 6, no. ½ (1992): 39–59.

Zhu, Tianwen. “Fin de Siècle Splendor.” In *The Columbia Anthology of Modern Chinese Literature*, edited by Joseph S. M. Lau and Howard Goldblatt, 388–402. Columbia University Press, 2007.

Chapter 9

"My Head Like a Mushroom Below"

Chen Kehua's Vegetal Gothic and the Anthropocene

Li-hsin Hsu

The intertwined ecological cycle between humans and plants has been a fertile ground for literary imagination across cultures. In the Chinese language, the two terms "matter" (物質/"wu zhi") and "plant" (植物/"zhi wu") not only share the same character "物" ("wu"/thing), but they are nearly homonymous, with the reversal of the term "物質" ("wu zhi"/matter) producing exactly the same pronunciation as "植物" ("zhi wu"/plant). This homonymous quality shared by the two terms might not be a coincidence. In recent years, there has been a materialist turn in eco-criticism, and plants occupy a crucial position in such an object-oriented eco-critical discourse.[1] In *Plant Horror*, Dawn Keetley points out the perceived monstrosity of vegetation that it "marks humans' dread of the 'wildness' of vegetal nature—its untameability, its pointless excess, its uncontrollable growth. Plants embody an inscrutable silence, an implacable strangeness, which human culture has, from the beginning, set out to tame" (1). The excessive overgrowth of plants, along with their soundless materiality, challenges existing aesthetic and social boundaries in literature, exposing often overlooked ecological associations among species. In the Gothic text, in particular, the forest is often seen to be infested with ferocious beasts.[2] For early pioneers, forests were unfathomable spaces, a savage environment unsuitable for human habitation, and thus connected with monsters and elves in cultural imaginaries. Among the various reasons for the fear of the forest, Elizabeth Parker notes that "Our fears of being lost to the woods are confused with fears of being consumed by the woods. The idea of disappearing into this environment is one that is evocative of being swallowed" (55). The vegetal ecosystem is frequently described as predatory, or even cannibalistic in popular imagination. On the other hand, plants also have this intimate biological relationship with all living things. As Keetley remarks, "while humans may occasionally become food for predatory animals, we all, whether buried in the ground or scattered on the earth, become sustenance for plants. Ashes to ashes. Flesh to food" (2016, 1). The symbiotic relationship between plants and all the other beings, their interconnectedness with humans and animals through the process of decomposition, dismantles the anthropocentric intent to dominate the ecological food chain. Indeed, recent critical attention has been paid toward the materiality of plants, and the trans-species interdependence that plants embody.[3]

Building upon previous scholarship, the chapter looks at three poems by an award-winning contemporary Taiwanese poet Chen Kehua (1961–)—"In a Thick Woods—On Effigy Mounts," "The Plant Empire," and "Plastic Flowers"—to rethink how his plant poems account for the enmeshed relationship between the human and the vegetal world. Ophthalmologist by vocation, Chen is known for his poetic engagement with social issues, particularly homosexuality-related topics. The recent English translation of his poetry collection *Guide for Garbage Classification*, nevertheless, shows a concerted effort to investigate the complexity of human-nonhuman entanglement through vegetative physicality and cross-species thinking. The chapter examines how Chen experiments with the materiality of vegetation in this collection to challenge preconceptions about what might be properly human, organic, or natural. The three poems, collected in the book, highlight the diverse material agencies of the vegetal world, their biological plasticity, their trans-scalar temporality, and their cross-species potentiality. Chen's vegetal imagination ranges from tropical woodlands to temperate forests and cemeteries, from ecophobic sentiment to anthropomorphism, from eco-spiritual mysticism to misanthropist environmentalism. Chen's "The Plant Empire" is further juxtaposed with "Song of the Chinese Juniper Forest" by Liu Kexiang (1957–), a natural writer well-reputed for his documentation of the environment in Taiwan. Diverging from Liu's poetic strategy of botanical lyricism, Chen's poems correspond with the contemporary anxiety about the Anthropocene by responding to the tension between pastoralism and modernity from an EcoGothic perspective. By probing his vegetal imagination about the materiality of plants (and its elasticity), the chapter showcases how Chen's vegetal Gothic destabilizes the human-nonhuman divide and speaks to the concerns of our time about multispecies sustainability in an age of the Anthropocene crisis.

NATIVE AMERICAN EFFIGIES AND THE SKULL-MUSHROOM GOTHIC DOUBLING

Among Chen's plant poems, "In a Thick Woods—On Effigy Mounts" demonstrates explicitly the unease felt by humans toward the uncontrollability of the nonhuman world. Written during Chen's participation in the Iowa Writer's Workshop in 2016, the poem depicts how the speaker visits the Effigy Mounds National Monument nearby and finds himself suddenly surrounded by a mysterious forest without a human trace. Both "the thick woods" and the "effigy mounts" in the title of the poem invoke a prehistorical sense of remoteness associated with ancient rituals, paganistic religion, and shamanistic practice of magic and sorcery. The locale of the Effigy Mounts specifically gestures towards the spiritual site used by premodern Native American mount builders thousands of years ago.[4] In such a natural-supernatural realm, the speaker becomes acutely aware of his isolation and fragility:

> Into the thick woods,
> One only sees mushrooms popping out from dead trunks of trees,
> No wildlife at all
> Anywhere,
>
> Except the hovering flocks of hawks overhead, observing
> My head like a mushroom below,
>
> Which can't help shiver a bit. (Chen 47, stanza I–III, lines 1–7)

In the midst of the wild woods, with "No wildlife at all / Anywhere" (line 3–4), the speaker "can't help shiver a bit" (line 7). Such a shiver implicitly evokes the Western literary convention of Native American haunting, a trope commonly used in the American Gothic and linked with white settlers' anxieties about trespassing former aboriginal territories. Renée L. Bergland comments on "the language of ghostliness" used by European American writers to describe Native Americans as a vanishing race (1). Such a "discourse of spectralization" plays a central role in shaping American national literature (21), since it serves as "a technique of removal" by replacing Native Americans "from the American lands" to "the American imagination" (4). Nevertheless, Native Americans, alive, dead or absent, continue to "haunt the white American imagination" (5). As a primordial archeological site, before the inception of Euro-American colonial expansionism, the Effigy Mounts conjure a wilderness that is both mythical and potentially sacrificial, with the "ghostliness" of the location prefiguring and foreshadowing the subsequent violence of colonial occupation and racial erasure through various Indian Removal acts and assimilation-related governmental policies. Allan Lloyd Smith highlights the four Indigenous features of the Gothic in nineteenth-century American literature distinctly different from those of the European counterparts: "the frontier, the Puritan legacy, race, and political utopianism" (163). Chen's poem, set in a potential hunting ground or ceremonial site of the Indigenous ancestors, speaks to such a layered discourse of spectrality, in which the racialized frontier space is entangled with the legacies of white settler colonialism, and the subsequent depletion of natural resources.

In this way, "In a Thick Woods—On Effigy Mounts" exerts a heightened sense of both human fragility and eco-precarity. What is removed or disposed of here, is not just the memories of Native American mount-builders, and the spirit-animals that those Indigenous mounts originally embody, but the disrupted or destructed ecosystem through the Anthropogenic process of modernization and settler colonialism. As Keetley and Sivils note, "relations of predation, edibility, and environmental exploitation have often been expressed specifically within the system of racial hierarchy and oppression" in American EcoGothic literature (8). Chen's poem conducts a similar ecological allegory in his vegetal Gothic here, which challenges a nineteenth-century mythical-historical sense of Manifest Destiny in a North American context. Chen's poem further relocates the realm of spectrality from the precolonial past to the Anthropocene present, by blurring the boundaries between the human and the nonhuman, among the vegetal, the human, and the animalistic.

In the poem, the decayed tree trunks, the mushroom, the hawk, and the back of the speaker's head, the most fragile and yet also exposed part of the human body, enact a multilayered entanglement among plants, animals, and humans, between race and the environment. Simon Estok notes that "Decay and rot are important to the ecoGothic because they are agency and excess overgrown and unpredictable. Literary treatments of rot and decay clearly reveal the ecophobic unconscious" (2019, 44). As Estok elaborates on this ecophobic nature embedded in the EcoGothic, "Rot and slime are unpredictable in their transgressions and blurring of borders and in their imagined alliance with an antagonistic nature. Corruption is the horror of uncontrolled agencies" (2019, 44–45). The "dead trunks of trees," along with the "mushrooms popping out" from the vegetal decay (line 2), serve such a transgressive border space that unsettles and problematizes human-nonhuman division as well as hierarchy. The sight of mushrooms feeding on or growing out of the "dead trunks of trees" invokes the biological process of life and death shared by all living beings (line 2). The connection drawn among the speaker's skull, the mushrooms, the trunk of the dead tree, and the hawks, creates a conflating and superimposing effect of the four organisms, imprinting on each other the relations of digestion, consumption, and decomposition. Furthermore, the parallels between the mushrooms feeding on the dead trunks, and the predation of the "hovering flocks of hawks / Overhead" over the speaker's skull, reveal the segue of the speaker's role from an innocent bystander and visitor, and perhaps an accidental trespasser and intruder, to an unwitting participant and an eventual victim in the bio-ecological practice of predation and consumption. In this way, the skull-fungi, head-trunk doubling perturbs the predator-prey echelons assumed by the human-speaker. While, as the speaker states in the poem, the human head resembles the "mushroom below," with the hawks in view, the speaker's ecological position is not unlike the "dead trunks of trees," prey for the bird as well as the tomahawk-like specter of the prehistorical hunting ground. In the forest without "wildlife" (line 3), the only human being, a visitor-intruder, might just become a sacrifice in this EcoGothic pecking order.

The eerie, uncomfortable atmosphere is deepened when the speaker in the poem notices the absence of wildlife in the preservation area. Such a lack of wildlife implies a potential environmental destruction linked with white settler colonialism, which turns prehistorical woodlands into a museum-like ecological graveyard. Chen ends the poem with an EcoGothic epiphany, with the skull-mushroom analogy subverting the predator/prey and victimizer/victim distinction. In a way, the poem visualizes a thrillingly dark vision of humanity as a co-participant in the Wicker-man-like endgame of human-nonhuman betrayal in this eco-historical haunting game. This mushroom-skull doubling is complicated by the liminal role fungi occupy in the triangulated relationship among humans, plants, and animals. Mushrooms reproduce and grow by hyphae and spores. Although their growing pattern is similar to that of plants, they cannot perform photosynthesis to produce chlorophyll. Anna Lowenhaupt Tsing thus notes that fungi are "closer to animals" because "[t]hey excrete digestive acids outside their bodies to break down their food into nutrients" (138). While, for Tsing, mushrooms are "world builders," since they can decompose substances into nutrients for other organisms (138), in Chen's poem, their world-making potentiality is more directly linked with the poetics of decay. In the Chinese version of the poem, the second word "顱"/"lu" in the term "tou lu"/"頭顱"/"head" in the Chinese language can denote both a head and a skull, and yet in the context of the poem, the term more specifically provokes a disturbing image of the skeleton or the bone structure of the head made visible in the process of bodily decomposition. The skull-fungi provocation brings to mind the Eastern practice of "white bone contemplation" ("baigu guan"/白骨觀) in Buddhism,[5] as well as the Shakespearean reference to the skull scene in *Hamlet*, in which the Danish prince contemplates upon the skull of Yorick, a jester in his father's court.[6] The parallel between skulls and mushrooms exceeds their ocular resemblance and evokes a complex cross-cultural, historical-ecological double consciousness, a Buddhist-Shakespearean moment of cosmic-philosophical lamentation about the impermanency of human bodily existence, compounded by an immediate, gothicized sensation of physical dissipation.

Chen's bio-poetic vegetality in "In a Thick Woods—On Effigy Mounts" illustrates what Eduardo Kohn calls the "living thoughts" in his 2013 book *How Forests Think*, and what Hubert Zapf perceives as "translation of natural into cultural energy" (Kohn, 16; Zapf, 163). Both critics draw upon Gregory Bateson's idea of the Mind-Nature analogy that associates metaphors in poetic language with "the principle of life" (Bates, 241), since analogies connect patterns across different living spheres. Chen's poem demonstrates the process of metaphor-making and vegetal-thinking by discerning structural similarities, shared properties, and emerging relations among species. The poem further suggests how the perverted "wildlife" cycle, as well as the spectrality of disappeared Indigenous practices, might have been the dire consequences of the Anthropocene. If the seemingly stomach-like mushroom will digest, or even swallow the speaker's skull like those dead trunks eventually, what might a post-human existence, rebuilt by those skeleton-mushrooms, look like? Chen's poem stops short of providing a glimpse of redemption from his skull-fungi analogy. The vegetal monstrosity like this exerts its haunting presence in these EcoGothic forests, feeding on the primordial psyche of human anxiety.[7] Not only does the forest at the Effigy Mounts instigate a deep sense of unease and fear in the speaker's confrontation with the nonhuman world, but the body (or skull) of the speaker is mapped onto the contour of the landscape, in the interfusion of which both the speaker's "humanity" and the mushroom's "nonhumanity" are disruptively called into question. In this way, the fungi-human Gothic Doubling in Chen gestures toward a post-human, vegetal-human-animalistic intimacy, with which Anthropos like the speaker, and by extension, us readers, have yet to reckon.

"THE PLANT EMPIRE" AND "PLASTIC FLOWERS"

In another poem by Chen, "The Plant Empire," the vegetal-human-animal entanglement further dismantles the speaker's anthropocentric epistemology. As the word "empire" in the title of the poem indicates, vegetal beings are the real dominant species of the earth, despite their peculiar quietude and motionlessness. Such an assertion might not be a hyperbole, since, as Stefano Mancuso and Alessandra Mancuso note, "The plant world alone represents more than 99.5 percent of the biomass of the planet" (123). And yet, the adverb "once" that begins the first line of the poem also accentuates the precarity of this vegetal claim to dominance in the face of human destruction:

Once, one plant-ruled empire was
Followed by another—
They had been so quiet,
And almost motionlessly
Ruling.

I, wind, clouds, the sky—
I'd met some trees that exposed the lung to the sky,
Who suffered ages of agony just
To breathe; I never had realized how
The annual formed

So introvertly, so slowly,
One circle formed beyond another,
Circling and circling into a solid
Stance.

I wonder how long that agonizing tear
between an exhale and
An inhale,
Would take—
I could only see things much briefer,
Such as blooming, such as shedding,

Such as mushrooms sprouting
from the fallen rotten trunks,
Such as weeds spreading over back yards
and paths.
But, I know I was heading into
This empire ruled by plants, many of
Whose names I did not know,
Many of whose species I could not tell—

Many of whom were guarding
a secret center. I knew
they were watching me quietly:
My animal body, animal heart,
Animal soul—
I knew I belonged to things
that lasted much shorter. (Chen 61 and 63, stanza I–VI, lines 1–35)

In the first five stanzas, the speaker adopts a speculative stance in his imagining of plants as a cult-like entity ("I knew / they were watching me quietly:," lines 30–31), echoing the representation of the secretive, observing wild woods in "In a Thick Woods—On Effigy Mounts." As is indicated by the enjambment of lines 5–6 with a stanza break in-between in the Chinese version,[8] "I, wind, clouds, the sky—" are the subordinates of those "quiet" and "motionless" plants. The alienness of the woody perennial species and its superiority is complicated by the trees' drastically different life patterns, whose annual tree rings are "Circling and circling into a solid / Stance" in the excruciating process of "breathing" (lines 9 and 13–14), through "that agonizing tear / between an exhale and / An inhale" (lines 15–17). The slow, "introvert," organic procedure of trees "that exposed the lung to the sky, / Who suffered ages of agony just / To breathe" (lines 7–9) forms a sharp contrast with the brief cycle of "blooming," "shedding," "sprouting," and "spreading" belonging to annual plants, short-lived vegetations, and non-vegetal beings (lines 20, 21 and 23), including floras, mushrooms, weeds, animals and humans. The speaker acknowledges his own epistemological shortsightedness and botanical blindness thus: "many of / Whose names I did not know, / Many of whose species I could not tell—" (lines 26–28). As the speaker confesses in stanza five, "I knew I belonged to things / that lasted much shorter" (lines 34–35). Chen's identification with other non-vegetal beings and self-reflectiveness ("My animal body, animal heart, / Animal soul—," lines 32–33) recalls the dominating position of plants from a geological point of view beyond human speculation.

It has been known that plants absorb carbon dioxide and release oxygen, a biological process fundamental to the survival of the human and nonhuman alike. Intriguingly, Chen in "The Plant Empire" projects such a crucial vegetal process of photosynthesis to be an excruciatingly painful one. Since the beathing and growing of annual tree rings are sensitive to the climatic conditions of each year, the photosynthetic process can be gradual and yet potentially disruptive for vegetal beings. Chen's depiction of the maturation of those trees draws one's attention to the dense materiality of the wood (in comparison with the softness of human flesh). The vegetal "otherness" of these plants ("They had been so quiet, / And almost motionlessly," lines 2–3) in the first stanza gains some understanding and sympathy in the next few stanzas, with their quality as sentient beings and capacity to suffer resembling those of humans and animals. By imagining their "lungs" expanding and contracting painstakingly to grow solid tree rings, the poem forms some sort of plant-human-animal alliance, bridging the human-nonhuman incommensurability temporarily ("I never had realized how . . . "; "I wonder how long . . . ," lines 9 and 15). The second half of the poem pushes such a human-nonhuman intimacy further by visualizing these trees as martyrs and guardians of all other beings "with utmost care":

> Walking past a garden of anonymous plants,
> I saw the plants was [were] using branches and leaves

To protect the morning dews with utmost care,
Keeping them so pristine and intact,
Hidden in the depths overlooked by the sunlight.

Indeed, each and every one was an oval diamond
And the diamonds belonged to things
much briefer than me;
I stepped on the plants,
Feeling how they snapped and fell
one after another,

Feeling, at the same time,

The resolution
Of guarding the humanity. (Chen 63, stanza VII–X, lines 36–49)

The second part of the poem presents a utopic vision of trans-species interdependence among all the residents in this "garden of anonymous plants," in which a sense of botanical egalitarianism is operated (line 36). The morning dews, like the walker-speaker, are equally protected and kept "so pristine and intact" by the "branches and leaves" of the trees despite their negligence by the sunlight ("Hidden in the depths overlooked by the sunlight," line 40). While like "an oval diamond," the morning dews don't last longer than the humans ("And the diamonds belonged to things / much briefer than me," lines 42–43), they receive "utmost care" from their botanical brethren equally. Nevertheless, such a bucolic, democratic vision (and perhaps a wishful thinking on the part of the human-speaker), is predicated upon the agony and hardship of the plants. The seemingly innocuous gesture of the speaker stepping on the plants causes their brittle branches to snap and fall "one after another" (line 47). It forms an informative contrast with the "resolution" of these botanical stewards to protect the human-speaker from harm (line 48). Their benign presence foils human-centric selfishness in the havoc the human-speaker causes. The "ruling" of the plant empire is not so much done through surveillance (line 5), as seems to be suggested in the first stanza, but mostly through its corporeal suffering, its sacrifice, and perseverance to safeguard the ecosystem.

The evocation of "mushrooms sprouting / from the fallen rotten trunks" in stanza five of "The Plant Empire" resonates with Chen's head-mushroom analogy in "In a Thick Woods—On Effigy Mounts," in which the fungi-skull Gothic doubling sends a shiver down the speaker's spine. In the poem, however, the speaker is more readily identifying himself with floras, mushrooms, weeds, and ephemeral plants with considerably shorter life cycles than trees. Such a trans-temporal comparison renders the human-vegetal division redundant. As the speaker repeatedly states that "I could only see things much briefer," and "I knew I belonged to things / that lasted much shorter" (lines 19 and 34–35), the vegetal dynasty and their longevity conjures up a geo-historical sense of consciousness that inspires humility in the human-speaker. The vigilant trees that seem intimidating, if not potentially antagonistic in the beginning of the poem, are shown to be profoundly caring and generous. The vegetal "secret" guarded by those trees in the first part of the poem is divulged in the second part of the poem to be their stewardship as the guardians of humanity ("Many of whom were guarding / a secret center," lines 29–30; "The resolution / Of guarding the humanity," lines 48–49). Chen's trans-scalar, trans-species imagination swerves towards a conservative direction at the end of the poem, in which the human-speaker's speculative stance, however utopic and democratic it may seem in a cross-species, trans-temporal sense, remains a human-centric fantasy. The "resolution" of the trees to guard humanity results mainly from the human-speaker's own projection, after all. And yet, by concluding the poem with the careless destruction (or atrocity) that the human-speaker inflicts upon the vegetal beings, the poem also highlights its articulation of an Anthropocene woe, which exposes the fragility and vulnerability of vegetal materiality, despite the apparent earthly dominance of plants in the past. Their imperturbability or imperviousness to elemental influences reveals their materialistic endurance, but as sentient beings, they also suffer and feel pain, just like animals and humans.

Chen's vegetal speculation magnifies the messiness of human-plant interaction, an aspect less explored in the works of contemporary environmental writings in Taiwan. To highlight the uniqueness of Chen's vegetal poetics, I juxtapose Chen's "The Plant Empire" with "Song of the Chinese Juniper Forest" (2001) by Liu Kexiang, a contemporary of Chen, who is known for his devotion to the conservation of Taiwanese mountains and vegetation. While both poems address the complexity of the vegetal ecosystem from an anti-anthropocentric perceptive, they employ diverging poetic strategies in terms of their speakers' positionalities. Both poets share their admiration for the interconnectedness among various living and nonliving beings in the forest, highlighting the essential role trees play in the biosphere. Similarly, both poems touch upon the motif of human insignificance and the ethics of care. Nevertheless, Chen's "The Plant Empire" is characterized by a tension between the human and the nonhuman, between human speculation about the otherness of vegetal materiality, and the trees' victimization under anthropogenic intrusion. Liu's "Song of the Chinese Juniper Forest," alternatively, portrays the "loftiness" of the Chinese Juniper Forest independent from and towering over humanity (line 22):

Finally, using a myriad of lofty postures
Together with the sky

They towered over the sea of trees

Moreover, separated by the length of a distant gaze
In a skewed world
They kept supporting one another

At that moment everything had matured
There were countless shades of green under the sea of trees
Undulating

Ferns clung to their chests
Groping in the direction of the sun's rays
Orchids clasped their arms
Revealing the glittering brilliance of the forest

Mountain clouds docked on their shoulders
Ready to drift away at dawn

Passing birds found lodgings in their hair
Anticipating the next day's journey

They are a community of elephants
Life inching them upward
Unrestrained for a thousand years
Looking to the depth and vastness of the sky

Looking upon their loftiness
We are travelers who fear solitude but will likely be faced
With solitude
In a fleeting instant
Incessantly gauging our own insignificance

Translated by Nick Kaldis (Liu 99–100, stanza I–VIII, lines 1–26)

Liu's poem begins by comparing the forest to "a community of elephants," separated from the "skewed" and presumably corrupted human world (line 18). Chen's poem, on the contrary, has the human-speaker supported and cared for, albeit in an inconspicuous manner, by the vegetal beings. In Liu's poem, humans are alienated from the Juniper Forest community as "travelers who fear solitude but will likely be faced / With solitude" (lines 23–24). For Chen, the plant empire once ruled the earth, including humans, with their unintrusive and attentive guardianship. Liu's poem depicts the Chinese

Juniper Forest as a dynamic vegetal universe on its own, with "the sea of trees" interacting intricately with the ferns, orchids, mountain clouds, and passing birds, the sky, as well as with each other (line 3). In Chen's poem, the plants are the custodians of humanity, as well as other beings, despite the human ignorance of their quiet distress. In Liu's poem, the vegetal community of the Chinese Juniper trees is dependent upon each other ("In a skewed world / They kept supporting one another," lines 5–6). The spectacle of "the glittering brilliance of the forest," while "separated by the length of a distant gaze" (lines 13 and 4), reminds the viewer of the meager, lonely existence of the human species.

The vegetal imagination in poems like "In a Thick Woods—On Effigy Mounts" and "The Plant Empire" by Chen and "Song of the Chinese Juniper Forest" by Liu provides a glimpse into the radical alternatives that botanical materiality might offer. Natania Meeker and Antónia Szabari comment on the "vegetal recalcitrance" that "Their way of being is material and nonanthropomorphic, yet profoundly generative for human thought and practice" (x and 1). The forest imageries in Chen and Liu speak to such a resistant vegetality and their generative possibility. However, Chen's poems show a more drastic poetic stance different from that of Liu in his (re)presenting and (re)imagining of the human-plant entanglement. Liu's aesthetic representation of the Chinese Juniper Forest reveals a detached, philosophical rumination upon botanical intelligence and elegance in the "loft postures" of the forest. Their regality comes across as a natural process of growth and communal solidarity ("Life inching them upward," line 19), a telling contrast with the human "fear of solitude" and alienation. Chen's poems present a more engaged speculative stance in the vegetal-human interaction, in which the plant-man boundary is precariously blurred, and often established upon a predator-prey relationship. In "In a Thick Woods—On Effigy Mounts," the human-plant intimacy is based upon a vegetal-animal-human cycle of edibility, consumption, and predation. "The Plant Empire," while emphasizing a human-vegetal empathy in the speaker's identification with the suffering of the trees, also exposes the human exploitation of the vegetal world. Chen's botanical imagination draws one's attention to the brittleness, delicateness, and frangibility of vegetal materiality, despite their elasticity and elemental recalcitrance. By doing so, Chen's poems call for a deeper level of cross-species involvement through the aesthetic swing from anthropomorphism (the tree-man empathy) to its reversal (the mushroom-skull doubling), swerving from an EcoGothic forest to a Botanical Republic, and then back to the Anthropocene sorrow about the human destruction of an earth that was "once" considered "the plant empire."

In a way, Chen's vegetal poetics stipulates a radical destabilization of human-centric epistemology. Kohn comments on the importance of thinking about "[H]ow other kinds of beings see us," since "encounters with other kinds of beings force us to recognize the fact that seeing, representing, and perhaps knowing, even thinking, are not exclusively human affairs" (1). Chen's vegetal speculation enacts such a crucial task of conjuring up various vegetal imageries to rethink the human-nonhuman relationship, particularly through their disjuncture, their subterraneous interconnectedness, and their uncanny doubleness. Chen takes on a misanthropist turn in his botanical speculation in "Plastic Flowers," in which the speaker encounters artificial plants in a cemetery on a wintery day in Boston:

> For these, I cannot forgive the human being.
>
> Yesterday, as I sauntered in the cemetery,
> There offered in front of the deceased's name on white marble
> Were plastic flowers,
>
> Whitish,
> Drooping in sadness,
>
> Like ghosts trapped in the mortal world:
> "Have you seen how the plastic flowers look like when they whither?"
>
> In comparison with the long-lasting, the rest just looks stale and old
> –
>
> And the human being will eventually
>
> Be gone a moment earlier.
>
> 2016,11,12 in Boston (Chen 71, stanza I–VI, lines 1–11)

Plastic flowers stand in here as some sort of manufactured duplicate of organic flowers, and yet for Chen, they are the plastic double reminding the speaker of what cannot be replaced. These "whitish" plastic flowers in the cemetery, "Drooping in sadness" (line 7), become the spectral figures here, with their disposable and yet non-biodegradable materiality symbolizing human mourning for the loss of their loved ones in a zombie-like state of immortality. If the invention and usage of plastic can be seen as an indication of modernity, plastic flowers then embody the polystyrene undead, with their hauntingly immortal presence "trapped in the mortal world" of loss, sadness, and death (lines 6 and 7). As the speaker asks: "Have you seen how the plastic flowers / Look like when they whither?" (lines 9–10). Placed in the context of the Anthropocene, the poem discloses how the human intent to pursue modernization and progress into perpetuity backfires (line 11), when what appears to be man-made and enduring pales in front of "the long-lasting" (line 11).

The term "the long-lasting" ("永在的"/"yong zai de") can refer to the perennial and regenerative, such as flora and vegetations, or some more abstract, idealistic and spiritual beings, things and ideas that transcend the physical and materialistic limitation of polyester objects. Ironically, as the manufacturers and consumers of polystyrene-related products, human beings "will eventually / Be gone a moment earlier" (lines 13–14), outlived by these drooping, withering artifices. In comparison with the vegetal dynasties in "The Plant Empire" that cradles humanity, artificial substitutes are corpse-like, short-lived, "look[ing] stale and old" (line 9). These plastic flowers are in sharp contrast with the living and breathing trees in "The Plant Empire" that, despite their agonizing growth, provide comfort, shelters, and protection to all the beings on the planet.

The Plastic-Plant Gothic doubling and the human-plastic-plant triangulation exemplified in "The Plastic Flower" showcase what Carola Hein calls "a global palimpsestic petroleumscape," a term used to describe "a layered physical and social landscape that reinforces itself over time through human action and connects urban and rural spaces, culture and nature, materials and intangible practices" (3). Plastics as a synthetic chemical material consist mainly of styrofoam made from crude petroleum and are closely linked to the fossil fuel industry. Turning genuine, organic flowers into plastic ones for their durability marks the capitalistic development of the manufacturing industry and mass production, and the modernity and urbanization associated with it. After all, in comparison with plastics, real flowers, like humans, "belonged to things / that lasted much shorter," as Chen states in "The Plant Empire." The plasticity of those flowers as materials and commodities in "Plastic Flowers" is emblematic of the urban replacement of living things and modern encroachment upon the nonhuman world with artificial objects produced in a petroleum-motivated economy in the process of industrialization. And yet, the plastic flowers, located in a cemetery, also bespeak the Anthropocene irony of the human longing for longevity, which results in the entrapment of humanity in various environmental crises, not the least the Great Pacific Garbage Patch, among many other toxins and pollutants. As the simulacra of the real things, those plastic flowers are destined to exist as some kind of post-human, Anthropocene ghosts that linger on the earth long after the human species is gone. As Chen begins "The Plastic Flowers" with a resentful declaration, "For these, I cannot forgive the human being" (line 1). The uncanny pairing of plants and plastic discloses the uncomfortable truth of anthropogenic aggression under the petroleum operation of the fossil regime.

CONCLUSION

The three vegetal poems of Chen remind the reader of the urgent need to re-examine the intimacy between plants and the human world through alternative ecological imaginaries. Chen's botanical speculation comes close to the concept of "double description" by Kohn, who utilizes Bateson's notion of the term to account for how human perception gains depth "[b]y recognizing the similarities and systematically comparing the differences between what each eye sees" (98; see also Bateson 64–65). As Kohn explains, thinking with forests through such double descriptions, which enables the brain to "compare[s] the differential duplication of ocular perspective," would "allow[s] us to see how we think like forests in ways that reveal some of the sylvan properties of the living thought itself as well as how we experience these properties" (99). Chen's poems, in a similar way, unveil the ghostly reality of eco-fragility in the Anthropocene through his ocular analogies and multiple perspectival shifts. In poems like "In a Thick Woods—On Effigy Mounts," "The Plant Empire," and "Plastic Flowers," the anthropocentric viewpoint is explictly decentered and reconstructed through the plasticity and elasticity of Chen's gothicized vegetal imagination. It switches from the ecophobic sentiment more commonly identifiable in EcoGothic literature in poems like "In a Thick Woods—On Effigy Mounts," to what I consider as the Anthropocene Gothic in poems such as "The Plant Empire" and "Plastic Flowers," in which the materiality of plants points toward the disturbing human-nonhuman interaction in the form of transgression. Vegetal beings, natural or man-made, serve as a witness to the victim/victimizer enmeshment between the human and the nonhuman world. Their potential alliance with humans and their betrayal posits as an emblem of a damaged ecosystem on the planet as a result of anthropogenic violence. The juxtaposition of the two latter poems, in particular, shows an ironic disparity between two hyperobject-like entities, forests versus artificial flowers, between vegetative organisms and plastic matter. The hetero-temporality of plants coexists with the trans-scalar plasticity of petroleum modernity. Timothy Morton calls human-manufactured objects like plastics hyperobjects that "are massively distributed in time and space relative to humans" (1). Chen's botanical imagination addresses such an uncannily multi-scalar, multi-layered human-nonhuman interconnectedness—the clash and disjuncture, superimposition, and triangulation—among various hyperobjective presences in the globalized petroleumscape. As Justin D. Edwards, Rune Graulund, and Johan Hoglund comment, "To live in the Anthropocene is to recognize that transgression, excess, and monstrosity are no longer anomalies in human life but inextricable parts of it" (xi). Chen's vegetal poetics, as seen in the three poems under study, exposes a monstrous potentiality of the intimate plant-man, vegetal-animal, plant-plastic embroilment. And yet, its speculative vigor also underpins the significance of imagining a vegetal-human utopia that might yet come true one day.

NOTES

1. A small section of the introduction was excerpted, translated, and adapted from the author's commentary piece, published in Chinese, in the CLAROC Newsletter no. 39 (2022).

2. Ecocritics have shown invested interest in exploring the theoretical potential of the Gothic mode in relation to environmental discourses. The emergence of Eco-Gothic is often linked to Simon Estok's 2009 essay on Ecophobia in "Theorising in a Space of Ambivalent Openness: Ecocriticism and Ecophobia," in which Estok emphasizes a deeper understanding and construction of nature through one's more frightening experiences with the nonhuman world. Tom Hillard's article "'Deep Into That Darkness Peering': An Essay on Gothic Nature" (2009) makes the association between ecophobia and Gothic literature explicit. In the 2013 collection of essays *Ecogothic*, Andrew Smith and William Hughes remark on how the Gothic's representation of "evil" in nature is often employed "for radical or reactionary ends" (2). Estok's *The Ecophobia Hypothesis* (2018) similarly explores the subversive aspect of the human-nonhuman interaction—the antagonism between humans and nature, or what he calls the "ecophobic condition," which "can embody fear, contempt, indifference, or lack of mindfulness (or some combination of these) toward the natural environment" (1). More recent publications, such as *Ecogothic in Nineteenth-Century American Literature* edited by Dawn Keetley and Matthew Wynn Sivils (2018), *The Gothic Nature* journal (2019–), Elizabeth Parker's *The Forest and the EcoGothic: The Deep Dark Woods in the Popular Imagination* (2020), and *Dark Scenes from Damaged Earth: The Gothic Anthropocene* (2022) edited by Justin D. Edwards, Rune Graulund, and Johan Hoglund, show an increasing critical interest in the relationship between aesthetics and environmental issues in the field of literary studies, and the continuous scholarly emphasis on their interconnectedness.

3. Michael Marde's *Plant-Thinking: A Philosophy of Vegetal Life* (2013) and the subsequent critical plant studies series published by Brill, Lexington, Peter Lang, and the like testify to the intellectual vigor of such a critical attention toward plant studies in recent years. In 2020–2023, there was a series of Critical Plant Studies reading seminars in Taiwan, led by Professor Liang Iping, President of the Taiwan Charter of the Association for the Study of Literature and Environment (ASLE-Taiwan). The reading seminar series was sponsored by the National Science and Technology Council (NSTC), and was devoted to the interdisciplinary studies of plants and various eco-critical thinking. The chapter is a result from such a collaborative endeavor.

4. The National Park Service website describes the Effigy Mounds as "ceremonial and sacred sites" built during the Late Woodland Period of the pre-Columbian North

America (1400–750 B.P.) by local Native American tribes ("Effigy Moundbuilders"). There are around 200 animal-shaped mounds preserved along the Mississippi River region in northeast Iowa and southwest Wisconsin, with the bear and the bird as two dominant animal mound shapes.

5. The "white bone meditation" is a traditional meditation technique related to the Buddhist practice of contemplation that visualizes one's body petrified or decayed into a corpse or skeleton ("buj jung guan"/不淨觀/Reflections on repulsiveness).

6. One of Chen's early poetry collections, titled *I Picked up a Skull* [我撿到一顆頭顱/"wo jian dao yi ke tou lu"] (1988), along with his later prose collection *Harvard.Hamlet* [哈佛．雷特/"ha fo . lei te"] (2003), show a similar textual reference to Shakespeare's *Hamlet*.

7. Keetley and Sivils provide three categories of what critics call the Eco-Gothic: "(1) a repository of deep unease, fear, and even contempt as humans confront the natural world; (2) a literary mode that uses an implacable external 'wilderness' to call attention to the crisis in practices of representation; and (3) a terrain in which the contours of the body are mapped, contours that increasingly stray beyond the bounds of what might be considered properly 'human' (4)." While Keetley and Sivils' categorization refers more specifically to nineteenth-century American literature, the first and the third points are particularly relevant here.

8. There is not a period at the end of the fifth line "Ruling" in the Chinese version, and thus the enjambment explicitly indicates that the following line "I, wind, clouds, the sky—" (line 6) functions as the object of the verb in the previous line (line 5). The period is added in the English version, and hence the slight confusion.

BIBLIOGRAPHY

Bateson, Gregory. *Mind and Nature: A Necessary Unity*. Creskill, NJ: Hampton Press, 2002.

Bergland, Renée L. *The National Uncanny: Indian Ghosts and American Subjects*. Lebanon, NH: University Press of New England, 2000.

Chen, Kehua. *I Pick up a Skull* [我撿到一顆頭顱/"wo jian dao yi ke tou lu"]. Taipei: Rye Field, 2002.

Chen, Kehua. *Harvard.Hamlet* [哈佛．雷特/"ha fo . lei te"]. Taipei: Chiuko, 2003.

Chen, Kehua. *Guide for Garbage Classification* [垃圾分類說明/"le se fen lei shuo ming"]. Translated by Wei-Cheng Lin. Bookman, 2018.

Edney, Sue. *EcoGothic Gardens in the Long Nineteenth Century*. Manchester: Manchester University Press, 2020.

Edwards, Justin D., Rune Graulund, and Johan Hoglund, eds. *Dark Scenes from Damaged Earth: The Gothic Anthropocene.* Minneapolis: University of Minnesota Press, 2022.

"Effigy Moundbuilders." The National Park Service. Available at: https://www.nps.gov/efmo/learn/historyculture/effigy-moundbuilders.htm.

Estok, Simon C. "Theorising in a Space of Ambivalent Openness: Ecocriticism and Ecophobia." *ISLE: Interdisciplinary Studies in Literature and Environment* 16.2 (2009): 203–225. https://doi.org/10.1093/isle/isp010

———. *The Ecophobia Hypothesis*. New York: Routledge, 2018.

———. "Theorising the EcoGothic." *Gothic Nature* 1 (2019): 34–53.

Hein, Carola, ed. *Oil Spaces: Exploring the Global Petroleumscape*. New York: Routledge, 2021. https://doi.org/10.4324/9780367816049.

Hillard, Tom J. "'Deep Into That Darkness Peering': An Essay on Gothic Nature." *ISLE: Interdisciplinary Studies in Literature and Environment* 16.4 (2009): 685–695. https://doi.org/10.1093/isle/isp090.

Hsu, Li-hsin. "An EcoGothic Dialogue between Matters and Plants" (「物質」與「植物」詭譎的生態志異對話/"wu zhi" yu "zhi wu" gui jue de sheng tai zhi yi dui hua). CLAROC Newsletter no. 39 (May 2022): 1–12.

Keetley, Dawn and Angela Tenga, eds. *Plant Horror: Approaches to the Monstrous Vegetal in Fiction and Film*. London: Palgrave, 2016.

Keetley, Dawn and Matthew Wynn Sivils, eds. *Ecogothic in Nineteenth-Century American Literature*. New York: Routledge, 2018.

Kohn, Eduardo. *How Forests Think: Toward an Anthropology Beyond the Human*. Berkeley: University of California Press, 2013.

Liu, Kexiang. "Song of the Chinese Juniper Forest." Translated by Nick Kaldis; edited by Frank Stewart Frank, Arthur Sze, and Michelle Yeh. *Mercury Rising: Featuring Contemporary Poetry from Taiwan*. Honolulu: University of Hawaii Press, 2003, 99–100.

Maitland, Sara. *Gossip from the Forest: The Tangled Roots of Our Forests and Fairy Tales*. London: Granta Books, 2012.

Mancuso, Stefano, and Alessandra Viola. *Brilliant Green: The Surprising History and Science of Plant Intelligence*. Washington, DC: Island Press, 2015.

Marde, Michael. *Plant-Thinking: A Philosophy of Vegetal Life*. New York: Columbia University Press, 2013.

Meeker, Natania and Antónia Szabari. *Radical Botany: Plants and Speculative Fiction*. New York: Fordham University Press, 2020.

Morton, Timothy. *Hyperobjects: Philosophy and Ecology after the End of the World*. Minneapolis: University of Minnesota Press, 2013.

Parker, Elizabeth. *The Forest and the EcoGothic: The Deep Dark Woods in the Popular Imagination*. London: Palgrave, 2020.

Principe, David Del. "Introduction: The EcoGothic in the Long Nineteenth Century." *Gothic Studies* vol. 16, issue 1 (2014): 1–8.

Sandilands, Catriona. "Mulberry Intimacies and the Sweetness of Kinship." *Ecologies of Gender: Contemporary Nature Relations and the Nonhuman Turn*. Edited by Susanne Lettow and Sabine Nessel. New York: Routledge, 2022.

Smith, Allan Lloyd. "Nineteenth-Century American Gothic." *A New Companion to the Gothic*. Edited by David Punter. Oxford: Blackwell, 2012. 161–175.

Smith, Andrew, and William Hughes, eds. *Ecogothic*. Manchester: Manchester University Press, 2013.

Tsing, Anna Lowenhaupt. *The Mushroom at the End of the World on the Possibility of Life in Capitalist Ruins*. Princeton, NJ: Princeton University Press, 2015.
Zapf, Hubert. *Literature as Cultural Ecology*. London: Bloomsbury, 2016.

Chapter 10

Reclaiming the Commons

An Ambient Poetics in Jessica J. Lee's Two Trees Make a Forest

Kathryn Yalan Chang

This chapter reads the "ambient poetics" in Jessica J. Lee's memoir *Two Trees Make a Forest* and analyzes how she is able to reclaim a common in an environment that is foreign to her.[1] By drawing on Timothy Morton's theory in *Ecology without Nature* (2007), I argue that like Morton's disavowal of the Romantic adoration of "nature," Lee's relation to the unfamiliar environment of Taiwan follows a similar trajectory. Disillusioned by the romanticized imagery of the island instilled by her mother, Lee experiences an "ambient poetics" in Taiwan, which is tantalizingly familiar and alienating at the same time. It is only through her imagination of a "planetary citizenship" where she is able to reclaim a common as Joni Adamson and Kimberly N. Ruffin advocate. I will start with a theoretical contextualization of Morton's work to help situate Lee's "ambient poetics" in the strange environment of Taiwan. Then I will apply the notion of "planetary citizen" to locate a commonality between Lee and the island.

"WHY AMBIENT POETICS?"

In his essay "Why Ambient Poetics? Outline for a Depthless Ecology," Timothy Morton asks, besides expressing sheer anguish, "what state of consciousness" would be appropriate for an era dominated by global warming?[2] With thoughtful introspection, he contemplates the possibility that a poetic framework can shed light upon this intricate state of consciousness, firmly asserting his conviction that such a framework exists aptly labeled "ambience."[3] He then defines "ambience" as "to make strange the idea of environment" in *Ecology without Nature: Rethinking Environmental Aesthetics*.[4] Morton emphasizes the significance of space as a fundamental element for the existence of an environment. To develop environmental awareness, he thinks it is essential to recognize "space as more than just a vacuum" and actively engage in "taking note of, taking care of, one's world."[5]

In order to engage in taking account of or caring for one's environment, we are at a pivotal historical juncture where humans have the ability to embrace or endanger the planet, driving us to reevaluate notions such as "common humanity, common wealth, and common ground."[6] Inspired by William Boelhower's notion of the three "commons," Joni Adamson and Kimberly N. Ruffin point out that the health of the nonhuman realm and the recognition by humans that they belong to the "local, national, global, and ecological communities" form the basis of the "planetary commons."[7] By integrating an irresistible planetary perspective with the well-being of all species, Adamson and Ruffin bring attention to Philip J. Deloria's fundamental questions in his foreword of their co-edited volume *American Studies, Ecocriticism, and Citizenship: Thinking and Acting in Local and Global Commons* (2013). In a context in which "local knowledge and local politics" are influenced by "transnational flows of capital, goods, and people," can a person genuinely embody the role of "a planetary citizen" while simultaneously being subjected to the influence of nation-states on the local (and consequently global) affairs at the same time? How might these actions be manifested? How can one identify and comprehend "the various scales of citizenship" that are involved? What tools and approaches can be employed to actively participate in a conversation?[8]

Based on Lee's background as a British-Canadian-Taiwanese writer who has spent most of her life in Canada, her mother's background in Taiwan, and Lee's notable engagement in recognizing and appreciating the environments of Taiwan, this chapter contends that her memoir, *Two Trees Make a Forest: Travels among Taiwan's Mountains and Coasts in Search of My Family's Past* (2019), exemplifies a compelling vision of embodying the planetary citizenship, as Adamson and Ruffin have argued. Through her keen observations and active contributions to the genre of nature writing, Lee's work offers insights and perspectives that highlight the potential for a nonlocal individual to embrace a global perspective and comprehend the local environmental issues that transcend national boundaries.

Moreover, I argue that Lee's writing of Taiwan's mountains, waters, and trees evokes what Morton has considered as "ecomimesis"—a rhetorical strategy common in nature writing that "generates a fantasy of nature as a surrounding atmosphere."[9] By eschewing the use of "nature," which accentuates the nature/culture divide and potentially reinforces the notion of superiority, Morton introduces the concept of "ecomimesis" as an alternative, equating it to a rough Greek translation for nature writing and emphasizing the act of imitating or mimicking the *oikos* (Greek for "home").[10] Rather than portraying where one is from, ecomimesis tends to show "where they are" instead; as a result, one "glimpses the environment rather than the person" in such writing.[11] Morton believes that ecomimesis involves invoking a specific place and time in which the writer is situated; writing that evokes "an environment" is considered as the ambience of ecomimesis, suggesting that detailed descriptions of the physical surroundings would help the writer create a vivid sense of place and engage the reader in the narrative world.[12] If ecomimesis is a literary technique that involves describing physical surroundings in great detail in order to convey a sense of environment or atmosphere, then, for Morton, "the fundamental properties of ambient poetics" are essential to the concept of ecomimesis.[13]

From this perspective, Lee's memoir exhibits a consonance with Morton's notion of "ambient poetics." Moreover, her exploration of Taiwan's nature-cultues[14] contributes to enriching Morton's idea of ecomimesis. Known for his critical analysis of the field of Romantic ecology, Morton challenges the authenticity of "the images of ambience" found in Romantic literature, the images deriving from diverse variations of "modern fantasy and figuration."[15] He believes that the artistic portrayals of the abstraction and vastness of space, as captured in "ambient poetry," "resides a freshness that is medicine to that poison."[16] In other words, Morton observes that there are two types of ambience, one of which could be an antidote to the other's detrimental influences. Ambience, possessing a dual element, could serve as a comforting remedy for "capitalist and technocratic alienation," ranging from the soothing music of the New Age to the masking of anguish in Beethoven's music at Auschwitz.[17] On the other hand, ambient images also allow for the entry into an alternative dimension that transcends conventional notions of "here"—such as nation, race, gender, as well as the idea—that universalism is a neutral medium through which all of these terms are held in place.[18] For Morton, "ambience exists beyond any notion of conceptuality."[19] Moreover, ambience as "dialectical images" both conceal and reveal the potential for a deeper perspective that challenges the self-centeredness that isolates humans from other life world.[20] Morton celebrates the role of ambience in communication. He indicates that ambient poetics draws attention to the "contact" or continuous dimension, the medium or environment in which communication occurs.[21]

By drawing on Morton's work, I argue that Lee's nature writing in her memoir embodies Morton's notion of "dialectical images" and his ambivalent stance toward "nature." Lee's work, through this lens, exhibits her profound appreciation for the interplay in Taiwan's naturecultures, while also demonstrating Morton's concept of "ambient poetics." Lee's memoir provides a valuable platform to explore various aspects of Taiwan, including its political turmoil, geographical composition, intricate cultural tapestry, diverse communities, historical narratives, and collective memories. Moreover, Lee's writing explores the myriad landscapes of Taiwan, encompassing its majestic mountain ranges, lush forests, winding rivers, expansive waters, and captivating coastal regions. Through an examination of these elements, I will illuminate the ways in which Lee's memoir contributes to a renewed understanding of the commons.

AMBIENT POETICS AS THE "REFLECTIONS OF ABSENCE"

Lee's memoir highlights political tensions and controversies surrounding Taiwan's status as a self-governing democratic country by likening Taiwan to a mysterious figure draped in "a veil of gray," unknown and "unrecognized by so much of the world, like many disputed places, its status is not a given."[22] Lee even compares Taiwan's controversial political position to the earthquake, emphasizing the condition of vulnerability, uncertainty, and unpredictability. As Lee states, "Names here are buried and written-over things, erupting from the ground underfoot the way faults emerge from a quake."[23] Lee's mother repeatedly accounts for the frequent occurrence of storms and earthquakes. As a result, Lee has developed "a certain macabre fascination" with "the turbulence of the island."[24] When Lee describes how the dangers permeate the island, including "earthquakes, landslides, coastal erosion, land subsidence, volcanic eruption," she suggests a sense of awe and admiration for the strength of the island, as well as a recognition of the need to take these perils seriously. Lee acknowledges that her grandfather's passing in Taiwan has fueled her curiosity and desire to comprehend the island he had considered as home. As a result, she is eager to investigate and gain a deeper understanding of Taiwan despite her many hesitations. It is these hesitations and reservations that make up the "ambience" at the beginning of her journey.

Morton's intention in using the term "ambience" is to challenge "a particular view of nature."[25] The term "ambience" is derived from the Latin ambo, which means "on both sides."[26] With its emphasis on "a sense of a circumambient, or surrounding, world,"[27] ambience "is not necessarily on the side of nature."[28] For Morton, pause, silence, or suspension all create the effect of ambience. For example, Morton uses "ambient music" to "undo the normal difference between foreground and background" and emphasize "environing properties."[29] In this regard, in addition to tangible objects and structures, the environment also encompasses an intangible, "palpable, yet ethereal and subtle" atmosphere. For Morton, ecomimesis's task is to convey "this sense of atmosphere."[30]

Morton's central concern in *Ecology without Nature*, according to Dana Philips, revolves around ecocriticism, which is a type of nature writing that fails to embrace doubt and theoretical exploration. Ecocriticism is also recognized as another manifestation of nature writing that the field claims to analyze, but celebrates exclusively.[31] As I have discussed in the previous section, the dual nature of ambient poetics has something to do with the use of "apophasis," which is often accompanied by "a self-reflexive twist."[32] This self-reflexivity contains "negative delineation" and both play important roles in ambient poetics. Morton contends that "negative ecology," which emphasizes what it is not, rather than what it is, is "a more apt description of nature."[33] However, as Morton suggests, this "negative poetics of the environment" will still suffer from limitations to accurately "describe something."[34] By containing "negative quantity" in their work, such as suspense, absence, silence, and loss, the artist/writer can create "a sense of sheer space"[35] or evoke the surrounding atmosphere or the environment in which they are situated.

In *Two Trees Make a Forest*, the use of negativity adds yet another layer of complexity to Lee's experience of absence and loss. With her grandmother's recordings and her grandfather's letters, Lee carries the emotional weight of her family's history and is prepared to "inhabit the places" as if they were calling her.[36] Nevertheless, as she acknowledges in an interview that *Two Trees Make a Forest* is a documentation of her personal journey of trying to find "a sense of belonging"; despite spending time in Taiwan, a feeling of displacement, a recurring theme throughout the book, never ceases (*Sine Theta Magazine*). Retreating to the forests, however, gives Lee "a sense of sheer space" to consider the environment of Taiwan. Lee's search for her maternal family lineage and travel to Taiwan was motivated by the sense of unease and shame over her limited knowledge about her deceased grandfather who had Alzheimer and passed away alone on the island. The loss of languages (Mandarin and Taiwanese), memory, and identity also contribute to Lee's desire to know about her cultural heritage and family history in Taiwan. As her maternal grandmother always complains that she "had been too foreign for [them], too Western" (103). The complicated political upheaval between Taiwan and China makes Gong (grandfather), Po (grandmother), Lee's mother, Lee's sister, and herself hesitate to regard themselves as Chinese because "we weren't from a China that exists any longer. . . . No single word can contain the movements that carried our story across waters, across continents."[37] And that is one of the reasons why she does not feel she belonged.

As an environmental historian, Lee attempts to find solace in the mountains and forests in Taiwan; nevertheless, the inclusion of negative elements is spotted in the writing of her experiences in Taiwan's naturecultures. She acknowledges,

> I retreated to the forests, but even there, once in a while, I felt unmoored. Thoughts of my grandparents came often as I walked the mountains, rhythm giving way to notions I dared not entertain in normal life. My stride worked itself into a charm, hypnotic, and the worst came to me. I hadn't done enough. Hadn't learned enough Mandarin, knew no Taiwanese at all, and hadn't taken enough of an interest.[38]

This self-reflexive emotion of loss and negation is triggered again when Lee detects "a loss in her [mother's] tone," mainly because she and her sister have limited knowledge of Chinese,[39] which creates a distance from their maternal heritage. Lee's writing about Taiwan evokes an atmosphere of inarticulation: "I saw in Taiwan something of the ways that places draw us in—and sometimes push us away again, and there grew in me an inarticulate longing."[40] Through comprehending her grandfather's letters and a few tape recordings of her grandmother's, Lee writes, "In Taiwan I seemed to carry their haunting inside, ready to inhabit the places I found."[41] The word "haunting" accentuates the intense and unsettling emotions that have permeated Lee's visit to Taiwan.

These heightened bodily sensations, suspension of time, or a plateau of experience are also present in Lee's memoir when she hides/hikes in the mountains and forests on the island. Lee's reflections of absence are nourished by her constant walking in the hills and by her developing "a love for these mountains and their forests, a need to return and return again."[42] Each trip, however, designates a "pause" in her life. Drawing on the insights of Romantic essayist Thomas De Quincey's experiments with psychic and physical intensity under opium influence, Morton emphasizes when explaining the concept of intensity the effect of pausing as more than just "a mere hiatus or stopping" but rather as "a staying-in-place endowed with its own intensity."[43] The unsettling feeling arises again when Lee's prior "romanticized" views of Taiwan, influenced by her mother, are dismantled upon her arrival in the country during adulthood. Without falling into "an idealized 'pristine' nature" perception,[44] Lee describes her struggles in climbing Taiwan's high mountains in great detail. Yearning for a fulfilling experience amid the falling rain and rising ridges, Lee presses on, and her hopes are dashed as she confronts frayed ropes and a chaotic forest strewn with boulders at the peak, leaving her "disappointed."[45] The peak itself offers little respite, as it is merely a tiny space amid the trees, lacking the grandeur or tranquility she hoped for.[46] Instead of seeking "some elemental wholeness, unity, or monism" with "nature,"[47] Lee's writing conveys a palpable sense of dislocation and the arduousness of her journey.

Lee's ecomimesis challenges the notion of a complete dissolution between the internal and external realms, as well as the subject-object dichotomy. As Morton clearly indicates, "*I am immersed in nature*" is not a mantra whose repetition brings about its content. Thinking so is wishful thinking, otherwise known as beautiful soul syndrome."[48] Lee's writing adeptly captures a discordant atmosphere and the absence of harmony or a utopian ideal, skillfully avoiding what Morton calls "beautiful soul syndrome." Moreover, her writing exhibits a reflexive tone as she engages in introspection and self-analysis, expressing doubts and questioning her choice to undertake the trial alone. In declaring, "My worlds exist in halves,"[49] Lee acknowledges she is experiencing displacement and alienation due to the linguistic barrier, which serves as a metaphor for her struggle to gain a sense of self in a foreign environment. Her writing also records physical and emotional discomfort, alluding to a broader understanding of unease and dissatisfaction. Being aware of a profound sense of absence, encompassing both the loss of language and identity, Lee grapples with her own limitations and seeks a sense of familiarity and connection that however proves elusive. She claims, "I want to know this place, to feel some sense of familiarity, but that is not simple, not an easy thing."[50] This exploration reflects a more profound yearning for meaning and belonging, which remains unfulfilled in the face of an indifferent environment.

All in all, in contrast to the Romantic notion of "immersion" as many British writers sought solace in nature as a means to escape rapid industrialization during the late eighteenth and early nineteenth centuries, Lee's intention to "retreat" to the forests so that she could gain some "salvation" from "nature" ultimately leads to frustration. As Vince Carducci indicates, "ecomimesis seeks refuge in a virtuosic ideal of selfhood."[51] Still, it is unable to evade the "corporeal and communal relations" we have with others in our shared environment.[52] The idea of nature in Lee's memoir is not "a set of objects separate from us,"[53] but rather an atmosphere or environment that encompasses all.

DARK ECOLOGY: THE TRAGEDY OF THE COMMONS

In *Dark Ecology*, as "a solution to beautiful soul syndrome,"[54] Morton formulates a concept of "depthless ecology" to critically examine and challenge the limitations of Deep Ecology. Instead of emphasizing "ecological sentiment" which often involves "considering guilt and reinforcing subject-object dualism," Morton proposes "the notion of ambience,"[55] especially "a melancholic ethics,"[56] and celebration of the "dark side" of the "interconnectedness" of "the ecosystem."[57] As he wittily poses a question, "Instead of whistling in the dark, insisting we're all part of Gaia, why not stay with the darkness?"[58] By highlighting the significance of "pollution, miasma, slime" and elements that possess a quality of shimmering, engulfing, and decay,[59] Morton criticizes ecocritics who tend to avoid the repulsive "leakiness of the world,"[60] instead favoring aesthetically "pretty or sublime pictures of nature."[61] In this section, I argue that Lee not only appreciates the "ambience" of Taiwan's dao, shan, shui, and lin but also fulfills the purpose of dark ecology to "love the disgusting, inert, and meaningless."[62] Taking Morton's concept of dark ecology into consideration, I maintain that Lee's memoir explores the unpleasant and slimy "underside of ecomimesis,"[63] making it evident that her work not only reinforces Morton's critique of idealistic "nature" but also exposes the tendency of this idealism to exclusively portray ecological experiences through the lens of "love and light."[64]

Lee's memoir depicting Taiwan's natural beauty manifests a nuanced perspective that goes beyond mere admiration, incorporating elements of ambivalence and irony to convey her emotional alienation from the environment. As Morton reminds us that "Dark ecology undermines the naturalness of the stories we tell about how we are involved in nature."[65] Lee's descriptions of Taiwan's naturecultures resonate with Morton's concept of dark ecology, thereby exemplifying the exploration of pollution, colonialism, and other dark aspects of ecology. Thus, Lee's engagement with Morton's dark ecology has illustrated the following discernible attributes.

Moreover, given our era's heightened environmental and political vulnerabilities, Lee, as a planetary citizen, explores the tragedy of the commons in Taiwan. Lee's portrayal of the ambience of Taiwan's naturecultures consistently intertwines its positive and negative sides. As Morton reminds us, "ecological criticism must politicize the aesthetic,"[66] by which he problematizes "a monolithic, reified, and rhetorically reinforced concept of 'nature.'"[67] Having been an avid hiker for years, Lee was struck by the lushness of Taiwan's mountains, but laments its sorrows through researching its colonial history. She notices Taiwan's hills "form a natural boundary" between urban cities and majestic mountains, adorned with a diverse array of flourishing trees such as "camphor laurels, peeling elms, charcoal trees, banyans, and sugar palms."[68] The hillsides come alive with abundant vibrant greens, enriched by the lush ferns, colossal taros, and the uplifting presence of Diplofatsia parasols reaching toward the sky,[69] creating a positively captivating atmosphere. Nevertheless, while Lee describes trekking through forests and over mountains, the negative ambience resulting from the colonial exploitations of Taiwan's forests is also conveyed in her writings. The events of land enclosure and forced labor that Indigenous communities endured, similar to those of North America, are noted by Lee. Lee discovers a few remnants from this turbulent past can still be seen at the Nenggao Trail, such as "informational placards, an aging kiln used by Japanese police, and suspension bridges built by Taiwan Power."[70]

In order to gain a deeper understanding of Taiwan's past, Lee reads documents relating to Taiwan's history and geography. For example, several works of nineteenth-century British geographers as well as Emma Jinhua Teng's *Taiwan's Imagined Geography: Chinese Colonial Travel Writing and Pictures (1683–1895)* help Lee imagine Taiwan while traveling. Lee's memoir illustrates the dark atmosphere created by the Qing Empire and later Chinese authorities exploiting Taiwan's resources during the colonial era. Through Teng, Lee is able to comprehend Yu Yonghe's travelogue, which exposes the imperial demand for sulfur and the troubling efforts of the Chinese colonial authorities to suppress the Indigenous population. Likewise, in Henry John Elwes's account, Lee finds that the massive cedar trees felled for timber emphasize the dominance of the forestry industry, as tensions arise between Japanese guides, Chinese, and Indigenous workers. In short, Lee's memoir sheds light on the history of the island's deforestation, reflecting the exploitative nature of colonial endeavors.

With her collage technique, Lee effectively portrays the fragile and uncertain ambience of the devastated environment by interspersing descriptions of dark environmental conditions with phrases highlighting its beauty. Lee demonstrates the island's vulnerability and transitory state affected by natural disasters and human-induced catastrophes. As Lee observes, Taiwan's rapid expansion of cypress and cedar wood exports during the 1970s was followed by increased regulations regarding environmental protection. For nearly a century, the forests had been viewed as a resource for the nation as it rapidly developed; efforts to reverse their extinction ultimately required "a change in perspective."[71] In the context of Taiwan's ongoing colonial circumstances, there is a need to reconceptualize the commons as a decolonial endeavor. Therefore, after Taiwan achieved democracy in 1991, the Forestry Act was passed, prohibiting logging. The forests now hold tourism value and are recognized for their role in fortifying hillsides against earthquakes and landslides. Despite these positive changes, Lee reveals that the island still faces challenges, including frequent landslides caused by factors such as human activities, climate change, and geological vulnerability.

Lee's exploration of the impact of capitalism on Taiwan's cypress and cedar forests, where intense conservation battles unfold despite a logging ban, evokes Morton's critique of "beautiful ecological souls" by examining the interplay between ecology and consumerism. As Morton indicates that "the beautiful soul" becomes entangled in "a mode of Romantic ecological consumerism" where the forest is transformed into "a shop window,"[72] and "the ambience of a shop window" is embraced as "the temple of nature."[73] Furthermore, Morton also notices "capitalist thinking, and capitalist machinery, actively 'disappear' the workers who operate it."[74] Due to its relationship to these automation processes, ambience, in this case, is indicative of capitalist alienation.[75] While Morton's stance on "environmental justice" may lack clarity and justification,[76] I argue that Lee's memoir serves as a fitting testament to Morton's observations on consumerism, capitalism, labor exploitation, and the ongoing struggles in Taiwan. In pursuit of profits, illegal tree poachers in Taiwan's high mountains put their lives at risk in treacherous terrain, driven by the allure of finding "an object of value."[77] Predominantly comprised of precarious workers, they represent the vulnerable individuals who bear the brunt of exploitation within the capitalist system. In response to poaching, forest rangers employ tactical measures and utilize DNA databases, but they too face significant threats. Lee further notices that derogatory nicknames such as "mountain rats" (山老鼠) and the use of cartoon imagery further perpetuate the dehumanization of these workers, highlighting the devaluation of labor perpetuated by capitalism. Lee's observations in her memoir align with recent news reports that reveal the precarious situation faced by illegal loggers, who must be not only vigilant of law enforcement but also wary of potential buyers associated with "armed syndicates."[78] The stark reality of this situation represented in Lee's ecomimesis underscores the complex dynamics at play, where profit maximization intersects with systemic exploitation, ultimately resulting in dire consequences for the environment.

Lee's writing about Taiwan's naturecultures is always intertwined with positive and negative ambience, which conveys a kind of "situatedness," a feature of strong ecomimesis for Morton, and a rhetoric that David Simpson connects to "the urgency of impending and 'threateningly nondiscriminatory' ecological peril."[79] As Lee notes that "these slopes were shrouded in trees"; her description of how she was encircled by clouds when she rose above them illustrates an element of awe and wonder, even a sense of mystery and enchantment. Her detailed observation of the tree line being significantly "higher than on the European mountains [she's] grown used to"[80] adds to the impression of astonishment and amazement. Nevertheless, as Lee later admits that she cannot help but find an underlying fragility characterized by landslides, earthquakes, erosion, and rockfalls "behind the apparent solidity of these mountains."[81] It is possible that the slopes of the island could crumble as a result of tectonic movement or a wet period.[82] Lee's reminder reflects Morton's position, "No one now thinks himself immune from radical threat."[83]

RECLAIMING THE COMMONS: REVERENCE FOR THE LIVELIHOOD SPACES

Instead of exclusively emphasizing human belonging and comfort in "nature," which Morton sees as "a form of monism" that ignores the diversity and difference revealed by perception, Morton promotes "an engaged ecocriticism" called "ecocritique" that acknowledges "nature" as "the inassimilable other."[84] In *Dark Ecology*, Morton claims, "To truly love nature would be to love what is non-identical with us . . . to love the thingness . . . the mute, objectified quality of the object, its radical non-identity. Nature is not a mirror of our mind."[85] In other words, we love "the thing as thing, not as person in disguise."[86]

Lee's ecomimesis demonstrates several characteristics of Morton's proposal of recognizing the "thing." Lee notes that the unique green hue enveloping Taiwan's hillsides resemble a watery landscape rather than the conventional land she spotted in other countries evokes images of "darkened waterweed" rather than trees. As she appreciates, "That verdant hue is unlike any other I know."[87] Lee's introspective contemplations on anthropocentrism consistently reveal a comparable acknowledgment of such distinctiveness and singularity. Lee states that despite the teaching of Western education to shield her from anthropocentrism: "to resist the idea of nature as for us alone, of a forest providing arboreal answers to very human predicaments."[88] Using self-reflexivity, Lee acknowledges her own inadequacy in fully resisting this mindset. She sees "in this mountain a mirror for [her] misunderstandings, as if in knowing its nature [she] might find a way to belong to this place."[89] In a sense, Lee's reflexivity recalls Morton's message of refusing to seek "some false oneness,"[90] but rather acknowledging "the gap is a paradoxical way of having more fidelity to things."[91]

Lee's memoir contains several passages that demonstrate neither the harmonious connection with nature nor "nature" as the passive matters to be dominated. As Lee felt frustrated while climbing the mountain, she sensed that "the forest closes around me. And it feels intentional."[92] While understanding "the forest offers no succor,"[93] Lee felt exhausted. The disturbance of leaves and the unsettling grunt from within the bamboo increased her anxiety. The unknown sound made her pulse race. A male territorial macaque's intimidating behavior, a watchful gaze, and aggressive grunts made her freeze, swiftly turn and escape. All of these signify not only the discord between the humans and nonhumans, but also the significance of fear in face of other species. Furthermore, despite her background as an environmental historian, which enables her to navigate Canadian pine forests and European heaths with ease, Lee finds herself "botanically adrift" when confronted with Taiwan's vast and unfamiliar woodlands. Taiwan has an abundance and diversity of plants that make identifying and naming them difficult for her.[94]

Although Lee candidly admits her frustration with recognizing Taiwan's plants, her writing suggests that colonizers also experienced a significant sense of fear in Taiwan's mountains and forests, particularly in the eastern parts of the island. However, she is unable to identify a resonating sentiment in the language of fear used by the colonizers. In Lee's view, the colonizers' accounts portrayed the interior regions of the island with "a trace of terror," with foreign vegetation and dense forests. Taiwan was depicted as both beautiful and menacing, an overgrown wilderness where the sublimity of towering mountains found its counterpart in the depths of "darkened woods."[95] Beauty, fear, curiosity, exoticism, and even disgust, were all woven into these depictions. As Lee acknowledges, "Though written in English, I struggled to find in them a language I could share."[96] Due to their fear of the wilderness, colonizers attempted to eradicate the mysterious and foreboding aspects of the forest. This aligns with Horkheimer and Adorno's critique of the Enlightenment in *Dialectic of Enlightenment*, where the first sentence states that Enlightenment has perpetually pursued the goal of "liberating human beings from fear and installing them as masters."[97] Enlightenment as well as positivistic scientific ideas led to the perception that nature is predictable and comprehensible. As a result, nature no longer appeared to be "dark, fearful, and mysterious," but was reduced to an object that could be tamed over time.[98]

A fear of nature may result in the total annihilation of the dark woods, causing disenchantment, demystification, and objectification of anything that appears to be suspicious to a modern mind. Morton's *Dark Ecology* is a valuable contribution in this regard. As Aretoulakis indicates that ecocritiqe and ecomimesis are two of Morton's concepts that aim to bring about "a re-enchanting, or re-mystification of the environment" and share the goal of "reassum[ing], or resum[ing], the mystery of the world" and a sense of wonder and reverence for the environment.[99] In a similar vein, following earth scientist and paleoclimatologist Andrew Glikson and primatologist and mammologist Colin Groves's proposal in their book *Climate, Fire and Human Evolution*, Dipesh Chakrabarty agrees that without restoring "a sense of *reverence* towards Earth," there is no indication that humans are capable of rising above "the realm of perceptions, dreams, myths, legends, and denials."[100] "Reverence" in Chakrabarty's research has a Latin root as "respect with fear, some kind of a feeling that this is much bigger than I am."[101]

A spiritual relationship with the earth is characterized by "reverence, wisdom, responsibility, [and] ethics,"[102] as Chakrabarty demonstrates, but these qualities do not equal Morton's condemnation of "those beautiful souls" looking for an "organist" or some type of "utopian one."[103] Though Chakrabarty and Morton initially appear to hold divergent perspectives regarding religion, their arguments ultimately converge when it comes to what constitutes an effective environmental awareness. Chakrabarty acknowledges that his discussions of reverence "are in the neighborhood of theology,"[104] which contrasts with Morton's belief that ecocritique challenges "religion and religiosity of all kinds," including "sincerity and guilt," as they perpetuate "a stable sense of independent and single identity beyond, above, or behind the world."[105] However, Chakrabarty clarifies that reverence encompasses not only "curiosity, wonderment, or biophilia" but also implies a relationship characterized by "respect mixed in with fear and awe," reflecting a sense of wariness.[106] As a matter of fact, Chakrabarty's notion aligns closely with Morton's concept of dark ecology, in which he recognizes that true ecocritical artwork rejects organic unity but "hold[s] the slimy in view,"[107] such as "a cyborg" or "a Frankenstein's monster."[108]

As Philips notes, Morton intends to expand and broaden ecocriticism, metaphorically likening it to opening up. However, the question remains as to how effectively Morton reconciles this metaphor of expansion and broadening with the less favorable imagery of the "entire nasty mess."[109] Within this framework, Lee's memoir embodies a synthesis of reverence and fear, entailing a combination of the beautiful and the ugly. As discussed previously, Lee's fear of other species is quite vital as Chakrabarty indicates that our existence within human civilization depends crucially on "not having to be fearful of most other life."[110] This absence of fear has become "a basic condition of modern life."[111] For Chakrabarty, a sense of reverence for the planet can be restored by fully appreciating its biodiversity.[112]

Several passages in Lee's memoir display elements of the Gothic and express how Lee exhibits reverence for Taiwan's commons in terms of their biodiversity. For example, in Lee's depiction, Shuiyang Senlin ("literally 'Forest of Overflowed Water,'") a forest formed by a landslide that blocked a stream near Mountain Ali in central Taiwan, holds a Gothic allure. The majestic thousand-year-old cypresses and cedars that grace the remote hills of Mountain Ali create an imposing scale. However, on September 21, 1999, the earthquake made the cedar forest transform into a haunting lake, with remnants of the wood standing like "ghosts" in the midst of "a still living landscape."[113] Lee and her friends, discouraged yet determined, embark on a seven-hour climb, battling with her fear of heights. The descent toward Shuiyang Senlin is treacherous, focusing them on navigating steep slopes and clinging to fixed ropes. Fatigued and frustrated, the final stretch reveals a clearing where stark sunlight illuminates the bleached, barkless trees immersed in water. The atmosphere becomes "otherworldly," akin to "a horror film," as the forest graveyard emits a haunted ambience. Camping on the opposite shore, Lee contemplates the Taiwanian cryptomerioides, a species associated with hidden parts and coffin making, finding the view at Shuiyang Senlin eerily reflective of its name, evoking a sense of loss and the macabre.

In addition to Shuiyang, other places in Lee's writing also exhibit both a beautiful and Gothic appearance. For example, when Lee visited Jioufen, a village on the hills of Keelung, northern Taiwan, with her friend Charlene, she recollected how they navigated the treacherous steps, the village of Jioufen emerges through "a veil of rain," "its lantern-lined alleyways" shrouded in a haunting ambience that inspired the ethereal world. Amid the pale lights twinkling in the gray, fog envelops the coastline, obscuring the borderland between land and sea. A winter storm engulfs them in the monsoon corridor, leaving them standing, embraced by an eerie cloud.

Qixing Mountain is another example that Lee shows the beautiful but fearful atmosphere. Enveloped in the heavy rain and wool-thick clouds, the mountains around Taipei exude a Gothic ambience. In Lee's description, Qixing Mountain, part of the Datun Volcanic Group, looms mysteriously amid the fog and downpour, emitting an "otherworldly and fierce" sound and "sulfurous stench" that amplifies its volcanic nature. This region's volcanic chasms and sulfur mines exude an eerie atmosphere, with rocks tinged in yellow-green hues. All in all, many parts of the book immerse the reader in a captivating Gothic atmosphere as she surrenders to the allure of "the mist-glazed surface of Shuiyang,"[114] bears a bone-chilling winter storm in the hills of Jioufen, and the taste of rain and the feel of cold wind in Qixing Mountain create a Gothic experience, while the obscured views and enigmatic surroundings enhance the sense of mystery and enchantment.

Also celebrating the enigmatic allure, as Morton highlights, the ecological perspective propels us to "bathe in the environing ocean of our surroundings," cultivating "a better ethical stance toward species and ecosystems."[115] However, this does not imply endorsing a passive reliance on the system to handle everything, accompanied by a "sitting back, relaxing" mindset.[116] In Morton's opinion, such a mindset is reminiscent of an attitude that merely admires Nature from a distance. This approach, akin to the effects of patriarchy on the portrayal of women, displays "a paradoxical act of sadistic admiration" (5). According to Morton, respecting other species and ecosystems involves "a choice"; that is, it demands that we make a conscious decision influenced by both "contingency" (because it is our decision) and "desire" (because we yearn for a change).[117] Holmes further explains that Morton's "new techniques of nature writing" employs "the contingency of natural reality without resolving it into a higher unity."[118] Without attempting for "a higher unity" with Nature, Lee's nature writing embodies "the contingency of natural reality"; as initially she harbored a negative impression of the Taiwanese wilderness.

Due to the lingering trauma inherited from her grandfather's death and the profound sense of "guilt and grief," Lee has long perceived Taiwan as a haunting and fearful place before visiting the island. Nevertheless, as Lee points out that Gong's "death and Po's have brought [her] new possibilities for knowing."[119] After getting chances to know the stories of Taiwan's common and empirical experiences, Lee discovers the weight of sadness is gradually alleviated. Through the cedar forest, she reveals a newfound connection. She finds this "a place where the old trees can span all our stories, where three human generations seem small."[120] Through being around the trees, Lee understands the sad stories of multiple generations appear diminutive, and the enduring presence of the forest remains steadfast, showcasing its remarkable biodiversity as a testament to its resilience. Lee admits that she has finally developed an affection for these mountains and forests, and a strong desire to revisit them.

As Morton notes, in early ecological science, the term "niche" was coined to reflect the concept of "home," deriving from a place that "houses a statue,"[121] and it served to describe the ecological role of a species within its environment by drawing parallels to how a statue finds its place. Morton asks, "Where does that leave migrating birds, hominids, pilgrims, gypsies, and Jews?"[122] As a person who looks for roots/routes in Taiwan, Lee answers such question to some extent. Roaming in Taiwan eventually changes Lee's perception of Taiwan. As Lee admits, throughout the landscape and Taiwan's commons, she is able to discern "an expression of this place" in which she has a deep connection between herself and their lives, one that extended beyond the death of her grandfather, beyond a past she was not fully able to comprehend.[123] Lee finds Taiwan serves as a diverse habitat for both "migrant and endemic species," exemplifying its role as a home or "niche" to both Indigenous and migrant organisms.[124] Lee particularly suggests that the connecting image of the Taiwan Strait, which once bridged Taiwan to Mainland when sea levels were lower many years ago, serves as a corridor for migrating birds, floating plants, and diasporic human groups. As Lee summarizes, "In Taiwan's plants, I saw both movement and change: species adapting to climate, to altitude, to soil."[125] Taiwan's exceptional biodiversity highlights its significance as a "niche" for diverse groups to thrive and demonstrates the continuous processes of adaptation and evolution in response to the surrounding environment.

CONCLUSION

This chapter explores Lee's memoir *Two Trees Make a Forest* to examine how a foreign individual can reclaim connections by rediscovering affection for

Taiwan's naturecultures. Drawing on the critical insights of Morton's ecomimesis and his concept of ambience poetics, it has discussed Morton's concept of "reflections of absence" to explore how Lee's lack of language, memory, and identity motivates her journey to Taiwan and her struggle with the feelings of ambience. I argue that Lee's work captures ambivalence and irony, revealing the complexities of her search. Furthermore, this chapter critically examines ecomimesis by intertwining Lee's search for her maternal family's roots/routes with the tragedy of the commons and the impact of colonialism on Taiwan's flora, fauna, and her family. Lastly, this chapter emphasizes Lee's writing through the lens of ambient poetics, focusing on her encounters with the more-than-human world and her recognition of the languages expressed by the mountains, waters, and trees on the island. Through this emphasis, I anticipate a renewed understanding of the commons, where the interconnectedness of humans and nonhumans, is acknowledged and appreciated for a more sustainable and inclusive future.

NOTES

1. This article is a partial fulfillment of my research project (MOST 111-2410-H-143-020-).
2. Timothy Morton, *Ecology without Nature: Rethinking Environmental Aesthetics* (Cambridge, MA, and London: Harvard University Press, 2007), 52.
3. Morton, *Ecology*, 52.
4. Morton, *Ecology*, 34.
5. Morton, "Why Ambient Poetics? Outline for a Depthless Ecology." *The Wordsworth Circle* 33, no. 1 (Winter, 2002): 54.
6. Boelhower, William. "Side-by-Side with You: The Common as Foundational Figure." Redefinitions of Citizenship and Revisions of Cosmopolitanism. Special issue of *Journal of Transnational American Studies* 3, no. 1 (2011): 47.
7. Joni Adamson and Kimberly N. Ruffin. "Introduction." In *American Studies, Ecocriticism, and Citizenship: Thinking and Acting in the Local and Global Commons*, edited by Joni Adamson and Kimberly N. Ruffin, 1–20 (New York: Routledge, 2013), 2.
8. Philip J. Deloria. "Foreword." *American Studies, Ecocriticism, and Citizenship: Thinking and Acting in the Local and Global Commons*, edited by Joni Adamson and Kimberly N. Ruff, xiii–xviii (New York: Routledge, 2013), xvii.
9. Morton, *Ecology*, 77.
10. Morton, *Ecology*, 54.
11. Morton, *Ecology*, 33.
12. Morton, *Ecology*, 33.
13. Morton, *Ecology*, 47.
14. Donna Haraway coined the term "natureculture" to describe the complex and intertwined relationship between nature and culture. In her book *The Companion*

Species Manifesto: Dogs, People, and Significant Otherness (2003), she challenged traditional dichotomies and proposed new approaches to understanding the interconnectedness of humans and nonhumans.

15. Morton, "Why," 54.
16. Morton, "Why," 55.
17. Morton, "Why," 52.
18. Morton, "Why," 52.
19. Morton, "Why," 52.
20. Morton, "Why," 52.
21. Morton, "Why," 52.
22. Jessica J. Lee. *Two Trees Make a Forest: Travels among Taiwan's Mountains and Coasts in Search of My Family's Past* (New York: Catapult, 2020), 42.
23. Lee, *Two*, 11.
24. Lee, *Two*, 43.
25. Morton, *Ecology*, 34.
26. Morton, *Ecology*, 34.
27. Morton, *Ecology*, 33.
28. Morton, *Ecology*, 33.
29. Morton, *Ecology*, 47.
30. Morton, *Ecology*, 34.
31. Morton, *Ecology*, 151.
32. Morton, *Ecology*, 45.
33. Morton, *Ecology*, 45.
34. Morton, *Ecology*, 45.
35. Morton, *Ecology*, 46.
36. Lee, *Two*, 174.
37. Lee, *Two*, 11.
38. Lee, *Two*, 103.
39. Lee, *Two*, 15.
40. Lee, *Two*, 15.
41. Lee, *Two*, 174.
42. Lee, *Two*, 16.
43. Morton, *Ecology*, 46.
44. Steve Keoni Holmes. "Wither Ecocriticism in the Era of the Hyperobject? A Review of Timothy Morton's *Ecology without Nature* and *the Ecological Thought. Journal of Ecocriticism* 4, no. 1 (January 2012): 57.
45. Lee, *Two*, 110.
46. Lee, *Two*, 110.
47. Norman Wirzba. "Book Review of Timothy Morton, *Ecology without Nature: Rethinking Environmental Aesthetics*." *JSRNC* 2, no. 4 (2008): 529.
48. Morton, *Ecology*, 182–183; italics in original.
49. Lee, *Two*, 108.
50. Lee, *Two*, 112.
51. Vince Carducci. *Ecocriticism, Ecomimesis, and the Romantic Roots of Modern Ethical Consumption. Literature Compass* 6, no. 3 (2009): 634.

52. Carducci, *Ecocriticism*, 634.
53. E. J. Shelton Expedition Cruising in Solander: An Application of Ecology without Nature (Thesis, Doctor of Philosophy) (University of Otago), 2012, ii.
54. Timothy Morton, Dark Ecology: For a Logic of Future Coexistence (New York: Columbia University Press, 2016), 187.
55. Morton, "Why," 54.
56. Morton, *Ecology*, 186.
57. Morton, *Ecology*, 184.
58. Morton, *Dark*, 187.
59. Morton, *Ecology*, 159.
60. Morton, *Ecology*, 159.
61. Morton, *Ecology*, 160.
62. Morton, *Ecology*, 195.
63. Morton, *Ecology*, 159–160.
64. Morton, *Ecology*, 198.
65. Morton, *Ecology*, 187.
66. Morton, *Ecology*, 205.
67. Ross Wilson. "*Ecology without Nature: Rethinking Environmental Aesthetics*: Book Reviews." *British Journal of Aesthetics* 48, no. 2 (2008): 239.
68. Lee, *Two*, 18.
69. Lee, *Two*, 18.
70. Lee, *Two*, 76.
71. Lee, *Two*, 240.
72. Morton, *Ecology*, 115.
73. Morton, *Ecology*, 115.
74. Morton, *Ecology*, 86.
75. Catherine Grimm. "Review of *Ecology without Nature*." *Keats-Shelley Journal* 58 (2009): 175.
76. For example, Fiskio points out that "Unfortunately, Morton's most extensive engagement is not with environmental justice literature and criticism but with what he labels 'conservative ecocriticism'" (111). Acknowledging that *Ecology without Nature* serves as the starting point for Morton's ecology theory, I share agreement with Fiskio's observations and assert that Lee's nature writing could potentially enhance this "environmental justice" aspect while preserving the atmosphere elements.
77. Morton, *Ecology*, 139.
78. Victoria Jen and Jalelah Abu Baker. "Taiwan's centuries-old trees under threat by 'mountain rats' who chop them down for illegal sale." *CNA (Channel News Asia)*. 18 May 2023. https://www.channelnewsasia.com/asia/illegal-logging-taiwan-mountain-rats-crackdown-3497701.
79. Morton, *Ecology*, 33.
80. Lee, *Two*, 24.
81. Lee, *Two*, 66.
82. Lee, *Two*, 67.
83. Morton, *Ecology*, 33.

84. Janet Fiskio. "Book Review Timothy Morton. *Ecology without Nature: Rethinking Environmental Aesthetics.*" *Environmental Philosophy* 5, no. 1 (Spring 2008): 110.
85. Morton, *Dark*, 185–186.
86. Morton, *Dark*, 196.
87. Lee, *Two*, 18.
88. Lee, *Two*, 108.
89. Lee, *Two*, 108.
90. Morton, *Ecology*, 141.
91. Morton, *Ecology*, 142.
92. Morton, *Ecology*, 108.
93. Morton, *Ecology*, 111.
94. Lee, *Two*, 17–18.
95. Lee, *Two*, 19.
96. Lee, *Two*, 19.
97. Max Horkheimer and Theodor W. Adorno. *Dialectic of Enlightenment: Philosophical Fragments*, edited by Gunzelin Schmid Noerr. Trans. Edmund Jephcott (Standford: Standford University Press, 2002), 1.
98. Emmanouil Aretoulakis. "Review of Timothy Morton's *Ecology without Nature: Rethinking Environmental Aesthetics*." *Synthesis: An Anglophone Journal of Comparative Literary Studies* 1 (2008): 73.
99. Morton, *Dark*, 73.
100. Dipesh Chakrabarty. *The Climate of History in the Planetary Age* (Chicago: University of Chicago Press, 2021), 175.
101. Chakrabarty, *The*, 216.
102. Chakrabarty, *The*, 198.
103. Dana Philips. "Reviewed of *Ecology without Nature: Rethinking Environmental Aesthetics* by Timothy Morton." *Oxford Literary Review* 32, no.1 (2010): 156.
104. Chakrabarty, *The*, 198.
105. Morton, "Why," 55.
106. Carducci, *Ecocriticism*, 641
107. Morton, *Ecology*, 159.
108. Carducci, *Ecocriticism*, 641.
109. Morton, *Ecology*, 5.
110. Chakrabarty, *The*, 200.
111. Chakrabarty, *The*, 200.
112. Chakrabarty, *The*, 217.
113. Lee, *Two*, 230.
114. Lee, *Two*, 247.
115. Morton, *Ecology*, 167.
116. Morton, *Ecology*, 167.
117. Morton, *Ecology*, 167.
118. Holmes, "Wither," 159.
119. Lee, *Two*, 249.
120. Lee, *Two*, 249.
121. Morton, *Ecology*, 98.

122. Morton, *Ecology*, 98.
123. Lee, *Two*, 16.
124. Lee, *Two*, 18.
125. Lee, *Two*, 18–19.

BIBLIOGRAPHY

Adamson, Joni and Kimberly N. Ruffin. "Introduction." In *American Studies, Ecocriticism, and Citizenship: Thinking and Acting in the Local and Global Commons*, edited by Joni Adamson and Kimberly N. Ruffin, 1–20. New York: Routledge, 2013.

Aretoulakis, Emmanouil. Review of Timothy Morton's Ecology without Nature: Rethinking Environmental Aesthetics. *Synthesis: An Anglophone Journal of Comparative Literary Studies* 1 (2008): 72–75.

Boelhower, William. "Side-by-Side with You: The Common as Foundational Figure." Redefinitions of Citizenship and Revisions of Cosmopolitanism. Special issue of *Journal of Transnational American Studies* 3, no.1 (2011): 47–53.

Carducci, Vince. *Ecocriticism, Ecomimesis, and the Romantic Roots of Modern Ethical Consumption. Literature Compass* 6, no. 3 (2009): 632–646.

Chakrabarty, Dipesh. *The Climate of History in the Planetary Age*. Chicago: University of Chicago Press, 2021.

Deloria, Philip J. "Foreword." *American Studies, Ecocriticism, and Citizenship: Thinking and Acting in the Local and Global Commons*, edited by Joni Adamson and Kimberly N. Ruff, xiii–xviii. New York: Routledge, 2013.

Fiskio, Janet. "Book Review Timothy Morton. *Ecology Without Nature: Rethinking Environmental Aesthetics*." *Environmental Philosophy* 5, no. 1 (Spring 2008): 109–111.

Glikson, Andrew Y. and Colin Groves. *Climate, Fire and Human Evolution: The Deep Time Dimensions of the Anthropocene*. London: Springer, 2016.

Grimm, Catherine. "Review of *Ecology without Nature*." *Keats-Shelley Journal* 58 (2009): 174–176.

Haraway, Donna. *The Companion Species Manifesto: Dogs, People, and Significant Otherness*. Chicago, IL: University Presses Marketing, 2003.

Holmes, Steve Keoni. "Wither Ecocriticism in the Era of the Hyperobject? A Review of Timothy Morton's *Ecology without Nature* and *The Ecological Thought*." *Journal of Ecocriticism* 4, no. 1 (January 2012): 57–61.

Horkheimer, Max and Theodor W. Adorno. *Dialectic of Enlightenment: Philosophical Fragments*, edited by Gunzelin Schmid Noerr. Trans. Edmund Jephcott. Standford: Standford University Press, 2002.

Jen, Victoria and Jalelah Abu Baker. "Taiwan's Centuries-Old Trees under Threat by 'Mountain Rats' Who Chop Them Down for Illegal Sale." *CNA (Channel News Asia)*. 18 May 2023. https://www.channelnewsasia.com/asia/illegal-logging-taiwan-mountain-rats-crackdown-3497701.

Lee, Jessica J. *Two Trees Make a Forest: Travels among Taiwan's Mountains and Coasts in Search of My Family's Past*. New York: Catapult, 2020.

Morton Timothy. "Why Ambient Poetics? Outline for a Depthless Ecology." *The Wordsworth Circle* 33, no.1 (Winter, 2002): 52–56.

Morton, Timothy. *Ecology without Nature: Rethinking Environmental Aesthetics*. Cambridge, MA, and London: Harvard University Press, 2007.

Morton Timothy. Dark Ecology: For a Logic of Future Coexistence. New York: Columbia University Press, 2016.

Philips, Dana. "Reviewed of *Ecology without Nature: Rethinking Environmental Aesthetics* by Timothy Morton." *Oxford Literary Review* 32, no.1 (2010): 151–161.

Shelton, E. J. Expedition Cruising in Solander: An Application of Ecology without Nature (Thesis, Doctor of Philosophy). University of Otago, 2012. Retrieved from http://hdl.handlc.net/10523/2218.

Wilson, Ross. "*Ecology without Nature: Rethinking Environmental Aesthetics*: Book Reviews." *British Journal of Aesthetics* 48, no. 2 (2008): 239–240.

Wirzba, Norman. "Book Review of Timothy Morton, *Ecology without Nature: Rethinking Environmental Aesthetics*." *JSRNC* 2, no. 4 (2008): 528–535.

Chapter 11

The Cultural Narrative of Tayal's Millet Ark

Yih-Ren Lin, Pagung Tomi, Hsinya Huang,
Chia-Hua Lin, and Ysanne Chen

This chapter focuses on the narrative of recovering millet cultivation in Taiwanese Indigenous cultural context, through which we will be able to find inspiration about adaptive mechanism in the face of climate change.[1] It is told through the global-local interaction concerning sustainability under the impact of climate change. The Millet Ark Initiative, including the authors and our team, exemplifies a social practice of collaboration between Indigenous people and the academia in Taiwan. It started from an international collaboration but is based on sincere support from local Tayal communities. Divided into five parts, this chapter includes "The Beginning," telling how we started the Millet Ark Initiative; "Tayal Foodscape and Hidden Adaptation Strategy," explaining the importance of traditional ecological tradition; "Shifting, Fire-Fallowing, and Intercropping Cultivation in the Millet Culture," telling the process of millet cultivation; "The Bio-Cultural Diversity of Millets," describing the botanical and cultural features of the species; and finally, "The Tayal Women's Role," illuminating the role of Tayal women and their bonding with the land. It is our hope that the chapter will shed light on the cultural narrative of the Millet Ark Initiative, thus engendering a landscape of life for both plants and humans.

THE BEGINNING

In 2013, Frederik van Oudenhoven, a Netherland environmental NGO organizer and a book-prize winner, got in touch with Yih-Ren Lin and Pagung

Tomi through the International Society of Ethnobiology's secretary, Natasha Duarte.[2] At the time, Oudenhoven was asked to establish a grassroots task force in response to climate change and in order to do so, he invited mountain-dwelling Indigenous farmers from different countries to discuss how climate change had affected their environment, as well as find a solution based on their local systems of Traditional Ecological Knowledge (TEK). This idea was founded on the presumption that the residents of the high mountains would sense the impact of climate change more acutely than those in the urban settings. For example, some of them had become victims—the so-called "climate refugees," like those in the mountains of the Himalayas or Tibet—due to the accelerated effect of the melting snow. Moreover, these Indigenous farmers who had resided long-term in the mountains must react to the ever-changing natural environment, so they have accumulated some local knowledge and created a culture around their adaptation to climate change. This was why we began to participate in the International Network of Mountain Indigenous Peoples (INMIP), a global grassroots network organization aiming to facilitate this kind of dialogue and cooperation. We anticipated the creation of a positive and proactive grassroots force and the discovery of a way to contribute as a collective voice from Indigenous peoples on the issue of climate change.

In the spring of 2014, this grassroots network concept finally took off. We and a few Taiwanese Indigenous people, including Apuu Kaaviana from the Kanakanavu tribe, and Atung Yupas and Yapit Tali from the Tayal tribe, were invited to Bhutan to meet with mountain farmers from nine other countries and participate in a walking workshop. This process and outcome were later broadcast by Taiwan's Indigenous Television Network. The walking workshop took us to the land of the Indigenous people in Bhutan, and climate change was the main point of discussion. As we walked and talked, we pulled together different experiences and opened up an opportunity for mutually beneficial sharing among different Indigenous peoples. During the five-day event, we came up with a formal and public declaration on our stance regarding climate change. Our representatives and the International Institute for Environment and Development (IIED) then presented it at the international conference on climate change. In addition, we took part in the 14th International Society of Ethnobiology conference held in Bhutan. More importantly, as an Indigenous woman from the remote mountain areas in Taiwan, it was Pagung's first time to hear from, through translation, the various farmers from around the globe who shared their experiences dealing with climate change. She was very moved and inspired, especially in terms of protecting traditional crops. This event later triggered her to reflect on what she could do from her own cultural background.

In 2016, the group of Taiwanese representatives, inspired by the spirit of walking workshop, launched the "International Indigenous Ecological Farmers Alliance Conference" in Taiwan, as a response to the Bhutan Declaration, through collaboration with the International Society of Ethnobiology. All the countries that attended the conference in Taiwan created the Farmers' Declaration together after the walking workshop. Then Taiwan's representatives, Apuu Kaaviana and Yih-Ren Lin, brought it up during the Democratic Progressive Party's Central Standing Committee meeting as a reminder to the elected party[3] of the Indigenous small farmers' agricultural policy, as well as their determination to respond to climate crisis. As one of the Indigenous representatives, Pagung decided to revert back to the traditional millet cultivation in her home village, Tbahu. In addition to the ecofarmers conference, her action was also influenced by the 2017 INMIP walking workshop in Peru, where Indigenous farmers shared their experiences as they toured the world heritage site "Potato Park," which showcased the Indigenous contribution of their local knowledge in conserving rich varieties of potatoes.

According to Pagung, millet is the shared traditional crop of the Taiwanese Indigenous peoples, and thus there is a deeply rooted connection between the cultural history and local knowledge among the Indigenous groups around the crop. The millet crop demonstrates not only biological diversity but also cultural and linguistic diversity. In other words, millet is not just the grain itself but also a connection to the land where the people belong to. Therefore, the key issue is the cultivation of the millet field and the related knowledge regarding millet's revitalization which includes water supply, soil fertility, and slope restriction. The fact that Taiwan is a subtropical Pacific Island with many natural disasters such as typhoons and earthquakes presages the high probability of Indigenous peoples' migration activities and forms their dynamic patterns of settlement living. Thus, the natural and social process of how the millet field is produced has to be dealt with seriously. As we began to engage ourselves with the social practices of millet field restoration, the whole traditional livelihood of the Tayal people started to unfold itself. This livelihood illustrates how the Tayal people's traditional migration, environmental change, and basic food demand have converged to the millet field cultivation, which we consider to be a foodscape production. The following are our observations and findings from the social practices and activism of the Millet Ark Initiative, including (1) Tayal foodscape and hidden adaptation strategies in their migratory history; (2) shifting, fire-fallow, and intercropping cultivation in the millet culture; (3) the bio-cultural diversity of millets in the context of Tayal livelihood; (4) the Tayal women's role and their connection to the land.

THE TAYAL FOODSCAPE AND HIDDEN ADAPTATION STRATEGIES

We will start our millet story from Tayal's traditional migratory practice. Our purpose is to understand the relationship between traditional food production and the ever-changing living environments. To the Tayal, migration is a common and communal activity, and more importantly, it is very possibly a result of the lack of food supplies under the impact of climate change. In this project, we have learned from the Tayal elders about their migratory routes for the past four hundred years. For the academic purpose of understanding "lmuhuw"—the traditional migration routes—different aspects of translation are needed, which include the recordings of the ancient singing, transcribing it into Tayal language, the translating activities to other languages such as Chinese and English, and finally the collaborative interpretation amongst the authors. The outcome is mainly presented in the form of a GIS map, which is to be viewed side by side with the lyrics and the accompanied interpretation. This is a way of deeper understanding of Tayal's Indigenous knowledge. Moreover, "lmuhuw" also contains the information about the Tayal spatial memory, ecological observations of their living environment, adaptive strategies, and historical migratory social action responding to the environmental change. In a sense, the migratory routes and their mapping embody Tayal people's foodscape, in which the landscape, such as forests, rivers, millet cultivation fields, and households, is perceived as a container of food production. It also illustrates how the changing environments affect Tayal people's livelihood. That is to say, migration is a strategy of adapting to the changes.

Traditional singing of "lmuhuw" is the best oral record of Tayal migratory history. It is Tayal people's narrative of their life, past, present, and future. The singing unfolds a deep culture of how the people respond to changes, which strongly links to their social institutions. Tayal is the biggest tribe in Taiwan. It is divided into several subgroups with distinctive linguistic differences. Each subgroup called "llyung" (literally, the group of one river basin) has its own singing which contains the migratory history of the group. In her fieldwork, Pagung heard the singing from the elder Yupas Pusing, recorded and translated it collaboratively with Atung Yupas and Watan Kahat. The Tbahu village belongs to the Mknazi subgroup of Tayal people. Seeking the supplies of food is always the central concern of the "lmuhuw" migration, which is made explicit in the following passages:

> Pin'ara ke nqu bnkis na Tayal lkotas ta lkmButa, lkmYaboh, lkmAyan
>
> (A reminder from lkmButa, lkmYaboh, lkmAyan, the three great ancestors of the Tayal people)

> laqi! maha ni si ta kaki squ qbuli na bnkis ta qani lga, musa ta pskyutan qu hwinuk ta la. mosa ta yan nqu qara na maqaw pqqolan nqu llaqi ta kinbahan la, mosa si psqquli mita wagiq la. nanu yasa qu usa ru usa hmkangiy kzikan mamu balung, usa hmkangiy 'san mamu matuk tnga na pazyeh, teta musa mbhuyaw pslabang mqyanux squ tgzyumuw na rhzyal qu laqi mamu kinbahan, teta simu minblaq mita squ sinnyaxan na wagi, mtasaw na qsya. yan na 'ali bzinuq mbhuyaw.
>
> (Children! If we keep on guarding our original homeland, we may starve and become thin. Descendants! Perhaps we may even fight over mountain peppercorn just to fill our bellies, and sigh to the skies. Thus, you must rise up, and seek hunting grounds that you can weave through, fertile land that you can plow. Let future generations flourish on bountiful land, accompanied by sunshine and clear rivers, and grow as quickly as a flying arrow).

This passage suggests the impact of environmental change that causes the shortage of food. Planning to migrate seems to be an alternative for the survival and thriving of the tribe. It also shows that the decision is made in a thoughtful way and with courage. Apparently, unlike the modern compulsory relocation policy determined by the government, Tayal people have the free will to carefully choose where they are going to stay and they do it with sophisticated preparations.

Furthermore, Tayal people migrate in solidarity and recognize themselves as oneness despite the coming split due to migration, as the next passage indicates:

> qutux lozi ga, laqi! usa ru usa blaq pgwayaw squ son mamu mtasaw nqu llyung, laxi usa mkkiy nqu qlcing, laxi usa pqeway na pakaw. kya qu qmayat squ tluqiy, ini ga blaq balay pgwayaw qu tunux na bubu, teta simu minblaq mbhuyaw.
>
> (Also, Children! Find a heart as clear as water, do not become cold and indifferent to each other. Do not cut off contact using thorns and fences. If there is someone you like, choose wisely and determine the bloodline, so you may thrive and grow in strength).

Some other "lmuhuw" passages show the Mknazi's migratory routes over the past 400 years. They identify Quri Sqabu, Papak Waqa, and Hbun Tunan as significant landmarks and as integral to the Tayal territory. In the Tayal creation story, "Pinsbkan" is their ancestral home, meaning "birthplace of the rock cracks." It is a metaphor for all the future generations moving north along the rivers and mountain ridges, expanding their territory, finding fertile lands, and dividing into various subgroups. In this chapter, we focus on how one of the subgroups Tbahu village's ancestors migrated to the group's current place as well as their geographical knowledge related to the environment.

Chief IkmButa is one of the three ancestral brothers and the direct ancestor of Pagung's villagers. In Tayal vocabulary, "quri" and "hbun" respectively mean the mountain saddle and the rivers' confluence. Both are the key landmarks for the Tayal to navigate the mountains. Hundreds of small place names describe their ecological observations along their migratory route. The observations embedded in the place names are evidence of their knowledge regarding how to adapt to the changing environment. This knowledge is vital to their safety and food supplies.

Moreover, any migratory action is a careful collective plan that includes many preparations in advance, such as the probing of possible residential sites and their land condition for future food sources. In "lmuhuw," as the phrase "so-called" implies that the route had been well explored by the heralds before the group set out on their journey. The Tayal remember their ancestral migration through "lmuhuw," which tells stories of not merely survival but the flourishing of the tribe. As diverse, healthy, and sustainable food and cultivation systems provide the needs of present and future generations, the Tayal people continue to thrive and flourish. It is worth noting that without the aid of modern technology, the Tayal ancestors have retained their ecological and geographical knowledge through singing their "lmuhuw." The reason for the migration is very much connected to the impact of environmental change. How to survive on a harsh migratory journey was strongly connected to the millet culture, as it is to be examined in the following section.

SHIFTING, FIRE-FALLOWING, AND INTERCROPPING CULTIVATION IN THE MILLET CULTURE

Tayal people have a myth about shooting the sun; this story is evidently the people's response to climate change in terms of natural history, which also has to do with millets. Legend says that a long time ago, two suns appeared in the sky and caused a great drought. The Tayal people chose brave warriors to carry infants on their back and go where the suns were. During the long journey, the children grew up and finally shot down one of the suns to restore life to Earth. Throughout the long journey, each warrior kept a tiny tube in his earlobe, and inside the tube were millet grains. Millets were to be sowed on their path, so that the warriors could find their way home and meanwhile secured enough food for their travels. The elders often repeat this story, as if to tell us that climate change is not a modern occurrence.

The millet plays an important role in the Tayal migratory epic. According to the elders, traditional millet cultivation has three important ecological secrets: first, fire-fallow cultivation; second, shifting cultivation; and third, intercropping. Millet is a drought-tolerant crop, so it does not need a complex irrigation system, and it is often grown on uneven lands and steep hills. The changes to agriculture due to limitations of the topography show Tayal people's adaptability to mountain life.

The Tayal practice fire-fallow cultivation because we do not separate forestry from agriculture as modern land management would. When the Tayal ancestors first arrived in a new location, they would clear up trees and weeds and then burn them. Next, they would mix ashes with the dirt and use them as a kind of natural fertilizer and start to farm. In the fire-fallow process, they used fire very delicately. Determining wind direction and when to ignite the fires are skills necessary for fire-fallowing; otherwise, it could lead to devastating forest fires. Through this farming method, the Tayal could artificially turn their woods into fields and, more interestingly, recover their fields into woods.

This use of land brings us to a discussion on shifting cultivation. Realizing how long-term farming would make the land less fertile and make it hard for the millets to grow, the Tayal practiced shifting cultivation. Unlike modern agriculture where farmers keep adding fertilizer, the Tayal ancestors chose to let their land rest. Letting the land rest is a special feature of the millet's shifting cultivation. Moreover, while they let the land rest, the Tayal grow Makino bamboo, which they call "luma" or they grow Formosan alder (*Alnus formosana* Makino), "iboh." The Tayal people are very dependent on Makino bamboo; bamboo shoots are a source of food, and bamboo stems are used for construction. As for alder trees, they are a heliotropic species that forms symbiotic relationships with the rhizobium that grows in its roots. Rhizobium has the ability to fix nitrogen into the soil, so once they burn the alder after a few years, the land retains the nitrogen-rich soil. Thus, through implementing shifting and fire-fallow cultivation, the Tayal people have created a complex adaptive ecosystem. They are a part of the environment, not apart from it. This could be what the North American Chief Seattle truly meant when he said that humans belong to the earth.

Intercropping is another special feature of the millet fields. Instead of saying that the millet is the only crop, you might say that it is the flagship crop behind the creation of the Tayal homeland. Traditionally, besides the millet, in a millet field there are also sweet potatoes, taros, corn, pigeon peas, and other different crops. Other grains and beans are grown among the millet, creating a multi-crop and nutritious diet for the Tayal, which contains protein-rich beans as well as starch. Furthermore, these crops also attract animals like birds, mice, and wild boars. The Tayal set traps for the animals in the fields, adding a new source of animal protein. The elders told us that in the past, millet was very rare and had to be carefully protected. One night, the person guarding the millet field heard a clanking sound and thought it was a thief. But upon listening carefully, he found that the noise was coming from beneath the ground and it was the sound of the sweet potato roots expanding. These roots created crevices and space in the soil, allowing the millet to sprout. The story of sweet potatoes helping the millet grow was a touching tale of ecological symbiosis. The concept of an ecological network has already long been passed down in the Indigenous culture. According to the Pacific Island Indigenous myth, Pea, Corn, and Cassava are three sisters. Cassava loosens the soil underground to let Corn grow vertically upwards, and Pea can grow along the cornstalk, creating an interdependent/ intercropping ecosystem. The intercropping tradition conveys strong ecological significance while the Green Revolution of modern agriculture endorses mass production of a single crop, going against the balance of nature.

In short, fire-fallow, shifting, and intercropping cultivation in the Tayal millet culture formulate a harmonious union of farming and the ecological system. This is most likely a result of long-term evolution and adaptation. Notably, this kind of inter-dependency and mutual support is critical under the context of climate change. As a major staple food, millets facilitate and shape the dynamic development of Tayal's traditional territory, in which the responsive strategies to climate change reside.

THE BIO-CULTURAL DIVERSITY OF MILLETS

According to Pagung, when she began to grow millets at her home village in 2016, she had little experience and received many skeptical eyes from the villagers. The village Tbahu like many other Indigenous villages in Taiwan has gone under the influence of market economy which means farmers got used to growing cash crops. Tbahu is a famous and an important site for growing tomatoes to supply the urban areas like Taipei. Few villagers would recognize the value of growing back the traditional crop, the millet, because they do not find it worthwhile in terms of income. Traditionally, millets are the staple food, but the millet plantations have been replaced by rice paddies since Japan colonized Taiwan in 1895, in order for the government to fix the Tayal people at one single location and prevent them from migrating. Therefore, the Tbahu village had abandoned millet plantations for quite a long time. Under such circumstances, Pagung's millet restoration was hard to realize; nonetheless, she managed to find millet seeds collected by researchers from other Tayal territories and preserved in the government's seed bank. She also found a piece of land and began to grow them.

Pagung received little attention until one day an old woman visited her millet field. She was surprised by her appearance and asked her why she was there. The old woman started to tell stories about millets, reminding Pagung of her childhood memories. The old woman began to tell Pagung knowledge related to millet including the long-lost vocabulary about the names of different millets, rituals, and agricultural practices to take care of the millet field. Pagung was so inspired that she started to interview the elders in the village, including her own father, and recorded their knowledge about millet. Ethno-classification of millets known by Tbahu villagers have enlightened her to the vocabulary of millet varieties.

Seven different varieties of millets have been identified since then, each carrying an Indigenous name, based on their colors and other characteristics. Indigenous peoples preserve millets by growing and caring them in their environment and everyday practices rather than in the seed bank as the mainstream bioscience does. In this way, language and culture of the millet keep biological diversity intact. Pagung has witnessed the whole process of recovering Tayal millets in the past eight years, at different stages of growing millet, for example, clearing the field by slashing and burning, seeding, weeding, thinning, bird-repelling, and harvesting. Her village people have started to plant the seeds by observing the blossoming of mountain cherry trees in February, but the phenology is now seriously challenged by climate change. Moreover, while slashing and burning practices used to be regarded by colonial Japanese experts as out-of-date practices, they have recovered sophisticated knowledge through the process, which we will explain in the migratory section. Rituals at different stages represent Indigenous people's relationship with the land and more broadly their conception of cosmology. Knowledge related to naming, resource usage technology and skills, social institution, spirituality, belief, and worldview are richly embodied through the Tayal's long-term practice of growing millet. The cultural and social life surrounding the millet would not have been possible if millet seeds were only preserved in the low-temperature refrigerator under the scientific project of the seed bank for biological diversity.

Pagung's initiative of recovering millets has finally gained support from a small grant from the Ministry of Culture in Taiwan, as well as collaboration from university partners. Over the years, one lot of land has grown up to ten plots and become a significant landscape of millets in the village. Attracted by the millet landscape, more villagers came to help as volunteers, especially older women. They started to tell stories about millets. Storytelling then becomes a common way of life when there are significant rituals held for millets such as seeding or harvesting. These stories include villagers' collective memory and knowledge related to millets. It also constitutes and reconnects the relations of the villagers to their traditions surrounding the land and ancestors.

THE TAYAL WOMEN'S ROLE

Millet culture is rapidly disappearing under policies that encourage the modernization and grounding of Taiwanese Indigenous peoples. In losing traditional cultivation culture, the biggest problem is the loss of the connection to the land and adaptability to environmental changes. When we joined the INMIP's walking workshop 10 years ago, we felt the importance of the grassroots force and traditional ecological knowledge to the climate change response. Thus, we started the Millet Ark Initiative. This movement was started by Pagung in the Tbahu village and is a preservation initiative regarding the bio-cultural diversity of the millet crop.

Taiwan has over 100 kinds of millet. Moreover, different villages have different traditional uses of millets. Millets can be steamed, made into congee or cakes, used for marinating meat or making wine; each village has its own methods. Pagung's village has not grown millets in the last 30 years, mostly because the need for modern currency caused villagers to switch to cash crops. The traditional phrases, rituals, skills, and social practices related to millets had all disappeared when farmers ceased to grow them. The preservation of millets means physically growing them on the fields, not resorting to any scientific methods including freezing seeds. Upon returning from Bhutan in 2014, our team began to focus on bringing back millet cultivation, especially in Pagung's village. She remembered that the seeds of different varieties were gathered from various Tayal villages and stored by the government. So, the first step was to get these seeds back and find the lot in the village to plant the millets. Thus, the millets grew from one field to over ten different fields over the past eight years. The number of participants has also increased from one person into an entire group of Indigenous cultivators.

Continuous support came from the elderly women in the village. They were the first villagers to approach the millet field, and they were the ones who started to share their memories of millet growing with Pagung. They shared many words related to millet, including the names of different millet types, the words for different stages of cultivation, songs, rituals for sowing, harvesting, and storing, as well as millet-related stories. The participation and memories of these elderly women added a lot of cultural and social meaning to the Millet Ark Initiative. Traditionally, the Tayal's patriarchal society demands women to be less vocal in public, but this does not mean that they are not proactive in the society. The millet preservation movement not only let us realize that women have an active role in the millet cultivation process, but also that a lot of the ecological knowledge is demonstrated in their everyday practices. More significantly, the efforts of the women have expanded the field of millets from one piece of land to a landscape of traditional food production.

The millet landscape is a soothing place, for it recalls the elderly women's cultural memories, which include many past feelings that were erased by modernization. Through physical labor in and contact with the culturally meaningful millet plantations, the process brings forth an intimate connection between the individual and community, between people and the land, and between people and the spiritual presence. This landscape is not "no human being's wilderness." On the contrary, the labor and participation of the Indigenous people enriches the land with an emotional connection, adding to the meaning of the "Tayal Country." The "Tayal Country" basically refers to the socio-ecological system created through the Tayal people's interactions with the environment. There are not just material but also biological and societal ecological services in the Tayal Country. This is the force that made the women willing to approach the millet field and gain a kind of healing. Our team members have participated in Tbahu village's sowing, harvesting, and storing process and simultaneously felt a kind of harmonious rhythm. The healing and effective impact of this millet initiative can never be overestimated. Unlike cultivating cash crops, millets bring smiles, laughter, and relaxation. This is because it is not done for monetary gain and these village women forge a sense of community and a bond of common livelihood through the childhood memories that they rediscovered and shared. They planted step by step on a terrace field, and the sowing work was finished quickly. As a consequence, the mothers went to rest in the shade, and someone started to hum traditional tunes of rimuy-rimuy-rimuy-so and voices soon joined to make a chorus. This is a soothing scene, in which re-growing the millet balances out the stress of work and distance between people and the earth caused by the logic of a capitalist market. Traditionally, millet growing is accompanied with the rhythm of seasonal change, the ecological interaction of the land, soils, and nonhuman species, the collaborative work across the age and gender of all family members, and the rituals to connect humans with nature as a whole. In our millet narratives, women take the lead to restore knowledge about millets and by extension, usher the community into the sustainable and resilient future through the present action of Millet Ark.

The millet story came to a symbolic performance when a food rights panel based on the Millet Ark Initiative was presented in the 2019 Annual Meeting of the American Studies Association in Honolulu, Hawai'i. The panel titled "Building Caring Solidarity Economies: Food Sovereignty, Community Solar, and Gastronomies of Place" conveyed Indigenous perspectives in building caring "solidarity economies" around food and energy systems, with Pagung being the primary voice of the Indigenous vision and wisdom. As a Tayal conservation initiative, the Millet Ark Initiative features Indigenous self-determination and place-making and brings to life the traditional foodscape which would otherwise be consigned into oblivion.

The international conference is not merely an intellectual event but metaphorically a long walk to be continued. This walk is about Indigenous knowledge and its relevance to climate change, from the ancient to the contemporary. Our team efforts in the last eight years had a chance to be presented internationally. The purpose of sharing our story was to explain our vision, which is a variation of the average conservationist's slogan, "Think Globally, Act Locally." Our activism emphasizes that the opposite—"Think Locally, Act Globally"—is just equally important. The global and local conservation joint efforts are mutually dialectical, and localized thinking shows the importance of traditional ecological knowledge. Our practical work in the last decade has reaffirmed the value of the ancient knowledge accumulated by Taiwanese Indigenous peoples with regard to the current global crisis. In the face of environmental changes, we must adapt and learn in accordance with the environment. Our Millet Ark Initiative delivers not merely the biological or cultural characteristics of the millets, but how millets-growing entails an understanding of the environment.

In Taiwan, millets are the flagship species of the Tayal people's traditional crops. Our experience exemplifies how the Indigenous people in the trans-Pacific context forge practices that invoke the concepts of Traditional Ecological Knowledge (TEK), Indigenous science, resilience, and foodscape. In one of the walking workshops held in the mountainous Tayal villages, Pagung sang, in Indigenous words, a song about the migratory journey of her Tayal ancestors, a group of whom walked across the sacred mountains of Papak Waqa and settled in Mknazi as "those who come first." It is in this historical process of migration and settlement that the Tayal people have developed ways of knowing and understanding the complex Austronesian world of islands, mountains, waters, and food, which the settlers from the Chinese continent only began to appreciate recently. This knowledge involves land management, relationship between humans and nonhumans, foodways, ceremonies and rituals, belief and practices, vocabulary and classification system, as well as the connection between the place and seasonal forces in our everyday life, which provide an intellectual stimulus to the project of Millet Ark Initiative. This updated chapter then provides an oral tradition of the Tayal journey—the "lmuhuw"—we have come along as a joint task force that has witnessed the millet restoration in the plantscapes of Taiwan.[4]

NOTES

1. This chapter is adapted from the essay, "Situating Indigenous Resilience: Climate Change and Tayal's 'Millet Ark' Action in Taiwan," by Yih-Ren Lin, Pagung Tomi, Hsinya Huang, Chia-Hua Lin, and Ysanne Chen, which was originally published in Sustainability, 2020, 12 (24), 10676; https://doi.org/10.3390/su122410676. We thank the journal publisher, as well as the essay authors, for the permission to include parts of the essay revised and included in this Lexington volume. The current chapter is a revision made in the style of the oral tradition.

2. The revision is conducted in the form of an oral tradition. It is mainly our collective storytelling of the journey led by Pagung Tomi, a Tayal millet farmer, and reconstructed by the joint efforts of the Millet Ark authors.

3. The Democratic Progressive Party (DPP) was elected to become the ruling party in 2016, which was the same year when the Millet Ark Initiative was gaining momentum.

4. As mentioned, this chapter is revised from our 2020 journal publication in *Sustainability*. We would like to thank the editor, Prof. Iping Liang, for the inclusion of our updated chapter in the Lexington collection; as well as the help and support of the Lexington series editors, Courtney Morales and Breasia Boyd.

Chapter 12

The Call of the Hyperobjects

Plants as Zones of Aesthetic Causality in Tao Lin's Taipei

Chingshun J. Sheu

Tao Lin's novel *Taipei* (2013) may seem to be an oddity in a collection of essays on critical plant studies in Taiwan: The novel has very few plants and is only tangentially about Taiwan. In my essay, "What We Talk about When We Talk about New Media: Digital Subjectivity and Tao Lin's *Taipei*," I have argued that the novel's main thematic concern revolves around the protagonist, Paul, and his navigation of a contemporary Westernized social landscape permeated by digital media and psychotropic substances.[1] There, my focus was on the various obvious and subtle ways that Paul's thinking and perception are molded to conform to digital, as opposed to analog, logic. This dominant theme ostensibly offers few openings for a reading based on critical plant studies. Yet the apparent lack of plants in the novel is precisely what forms the backbone of my argument here, in the sense that, if there truly is no place on Earth outside of an ecosystem—if we truly do not see nature as something *over there*—then every text with a realist component should lend itself to an ecological reading.[2] *Taipei* has a dearth of plants, not an absence of them. It also has a slightly more direct ecological connection through fungi, in that the novel's climax is fueled by a large dose of psilocybin mushrooms. And fungi, in the guise of mycorrhiza, play an indispensable role in the life of (land-based) plants. Though psilocybin-producing fungi are not significant participants in mycorrhizae, their growth is intimately tied to plants and so supports my main argument in the present study.

That argument, in brief, is that there is a specter haunting *Taipei*—the specter of the mycorrhizal hyperobject. To convey the full import of this statement, I will first locate the few mentions of plants in the novel, and then link them through the mycorrhizal network, or the "wood wide web," before introducing the notion of a hyperobject and sketching out how the mycorrhizal network constitutes a hyperobject to rival the novel's dominant hyperobject of the digital. The relationship between these two hyperobjects in *Taipei*, I argue, resembles the relation between writer Paul's identities as a Taiwanese American and as an online person. The successful tracing of this relation can gesture toward the possibility of foregrounding the novel's obscured mycorrhizal hyperobject, and the means sketched out thereby of transforming hauntology into a more flexible perspectivism can help us respond to large-scale phenomena such as global warming.

TREES AND MYCORRHIZAE

In the novel, among the plethora of observations perceived through digital logic, the numerous parties and gatherings, and the drug-fueled haze, there are also five mentions of trees, excluding artificial trees, imagined trees, and pictorial depictions of trees—I focus here on trees, as other land-based plants are mentioned twice each at most. Paul lives near "a shadowy, tree-heavy intersection."[3] In his elementary school days, a girl once "kissed [his] cheek in a tree."[4] Sitting in the back of an SUV, he observes, "The unlighted space, all around him—and, outside the SUV, the trees, sky—seemed more visible, by being blacker, or a higher resolution of blackness, almost silvery with detail, than normal."[5] Calvin, Paul's friend and a fan of his work, lives in an "upper-middle-class neighborhood" where "[m]ost front yards had one or two fashionably sculpted trees and two or more colorful Boy Scouts–like patches of flowers and plants in independent organization."[6] And Paul says to wife Erin of their marriage, "We'll be that tree" in five months, pointing to "a tree that appeared healthy and, he thought, dignified."[7] The first instance is the only one that has no overt symbolism, though we can extrapolate from the relatively extensive canopy in the middle of New York City that this particular intersection might have existed in its current state long enough to perhaps have some historical significance. The tree of the second instance connotes privacy for suburban children with stay-at-home mothers (else they would not be in the tree) and, coupled with the kiss, conveys a romanticized youthful innocence. The trees of the third instance, seen outside of the SUV, play (along with the sky) the traditional role of Nature as the (pure) Outside of human society. The stylized trees and independently organized flowers and plants of the fourth instance serve to illustrate Calvin's upper-middle-class background, and satirically gesture by extension toward the likely core demographic of the readers of Paul's writing. And the tree in the last scenario only catches Paul's attention because it seems to embody metaphorically the health and dignity of the "family tree" that he wishes for his brand-new marriage.

With the possible exception of the first instance, these examples invoke trees to express anthropocentric conceptions of vegetation. Though always already in the background of the text, trees are drawn into the narrative only when they are needed as symbols of "Nature." And yet we should remember that, in the realism of the story world, they are physically there, along with supposedly numerous other trees that go unmentioned. Their vegetal presence allows for the possibility of their interlinking, for the vast majority of land-based plants participate in the plant-fungus symbiosis known as *mycorrhiza*, as outlined by Merlin Sheldrake and Robert Macfarlane.[8] Plants need more than the products of photosynthesis—carbohydrates and protein—to lead healthy lives, and they get many of these other nutrients, such as nitrates and phosphates, from mycorrhizal fungi entangled with their roots; in return, the fungi receive carbohydrates and protein from the plant (though there are commensal and parasitic mycorrhizae, the predominant form is mutualistic). But this is just one plant–fungus pair. As most plants can symbiose with multiple fungi, and most mycorrhizal fungi can symbiose with multiple plants (and with other fungi), no one yet knows the size limits of these mycorrhizal networks, giving rise to their colloquial name: the "wood wide web." The wood wide web links plant with plant, enabling them to share and exchange chemicals. When nutrients are distributed among them suboptimally, the fungus redistributes them. And when one plant gives off chemicals to repel dangers, such as an aphid attack, the fungi spread those, too. It is in the fungi's interest to ensure the survival of the plants that provide it with nutrients. In many ways, then, it makes sense to view the participants of a mycorrhizal network as a single entity. The metaphor of the internet seems even more apt when we consider that mycorrhizal networks also transmit spikes of electricity whose frequency one researcher has compared to human language.[9] We might even call the "wood wide web" a hyperobject.

A MODIFIED CONCEPTION OF HYPEROBJECTS

To understand hyperobjects, we first must comprehend the meaning of *object*. I use this word in the sense given to it by a new school of philosophy known as object-oriented ontology. Object-oriented ontology is a "flat ontology," which in this case means that it recognizes only one type of existent: objects.[10] This is in contrast to traditional ontological divisions such as substance/accident, potential/actual, virtual/actual, subject/object, or object/relation; and it derives from object-oriented ontology's axiomatic belief that "all objects . . . equally exist while they do not exist equally."[11] That is to say, every existent need not be the exact same as every other existent, but they are all granted the same ontological standing, with no type of object taking ontological precedence over any other type. Moreover, there is more to an object than presents itself at any given time. We can never fully know an object; it always has the capacity to surprise us. Jean-Paul Sartre calls this "the transphenomenality of being";[12] Timothy Morton calls it the "strange stranger": objects are "strangely strange. Their strangeness itself is strange. We can never absolutely figure them out. If we could, then all we would have is a ready-made box to put them in, and we would just be looking at the box, not at the strange strangers."[13] Vinegar can be a cleaner. A dog can be a police officer. A Norton anthology can be a doorstop. To explain this, father of object-oriented ontology Graham Harman posits that each object comprises two parts, a sensual object and a real object, with each part granted its own qualities. Among these four elements are ten possible relations, including between an element and itself, and these four elements and ten relations can explain any phenomenon.[14]

This arrangement strikes me as oddly complex for a purportedly flat ontology, not to mention the confusing equivocation of the term "object." I propose a modified object-oriented ontology that simplifies matters significantly, a proposal that I have delineated in more detail elsewhere.[15] Rather than a ten-way fourfold, I posit a unified object with a perceptible and an imperceptible *side* (I use this term metaphorically). Whenever an object presents a new facet, we can say that the facet, previously located on the object's imperceptible side, is now on its perceptible side. *Object*, meaning existent, is defined broadly. Objects include medium-sized dry goods, and also unicorns, justice, both eating and having one's cake, and so on. The subject, as traditionally understood, is also an object, with the imperceptible side playing the role of the locus of subjectivity. This is one of the primary advantages of a flat ontology: that humans are, ontologically speaking, nothing special. We need not *understand* the subjectivity of a nonhuman object to *grant* it. Since every subject is an object, and there are only objects, then by syllogism every object is a subject; the term *subject* is rendered superfluous.

Objects are made up of other objects, and an object can, while still being its own object, be part of another object, depending on the perspective. A Norton anthology is a doorstop when viewed alone, but a textbook when the owner is studying it for a course—even if the book is still placed at the door. *Perspective* here does not presume a perceiver; or, to put it another way, the perceiver is an object, and not necessarily a human one. The Norton is a doorstop or textbook depending on how the owner is using it at the moment, yes, but a bottle of vinegar is either an ingredient or a cleaner depending on what it comes into contact with. It is a question of object relations, and one object relates, or *attunes*, to another through the latter's *zone of aesthetic causality*, so-called because perception is the minimal relation.[16] That object A attunes to object B through the *latter*'s zone of aesthetic causality is the same as saying that object B *tunes* object A through the *former*'s zone of aesthetic causality. As everything is an object, these zones are objects, too. For the Norton, the zone is the inertia of the door, or the gaze of the reader; for the vinegar, it is the chemical composition of the molecules with which it interacts. Every attunement causes a new object to be formed, comprising all the participants of the attunement[17]—the stopped door, the student of literature; the delicious dish, the spotless sink. By understanding relations in terms of the objects involved and the objects produced, this modified object-oriented ontology avoids the object/relation binary and retains its ontological flatness.

A *hyperobject*, then, is a type of object in this sense, one that is "massively distributed in time and space relative to humans."[18] It should be immediately apparent that this is a relative definition. We speak of humans because we *are* humans, and inevitably we view things from a human perspective; but in any attunement in which at least one object is vastly larger and more diffused through time and space than the others, the much larger objects are also hyperobjects. Just as society is a hyperobject for us, we are a hyperobject for our gut microbes. Hyperobjects should be considered their own type of object because of their relative spatiotemporal diffusion; we seldom notice one until faced with its zones of aesthetic causality. We are always part of a society, but we seldom notice until confronted with a piece of conventional wisdom, or until we are put into a situation where we must interact with strangers. Or, to return to our key example, most land-based plants are part of mycorrhizal networks, but we seldom notice this until, for example, we harvest and remove a key plant from its network, and it (and/or its network) suffers. Hyperobjects differ from other objects in this way. An object is defined not just by its parts but, more importantly, by how those parts are organized,[19] and because a hyperobject is so relatively massive, the organization of most of the new objects formed from attunement through its zones of aesthetic causality differs little enough from the original hyperobject (though they do still differ) that we can, for convenience's sake, consider the hyperobject unchanged. Adding a new person to a society usually changes the society as much as adding a new tree to a mycorrhizal network changes the network—very little. In the aforementioned instances when Paul notices trees, he is attuning to the mycorrhizal hyperobject and thus forming new objects with it—the leafy intersection plus Paul, the childhood tree plus Paul and the girl—but the mycorrhizal hyperobject itself is almost entirely unchanged.

Though Paul is almost completely insignificant from the perspective of the mycorrhizal hyperobject, the reverse is untrue. Each presentation of the mycorrhizal hyperobject significantly alters Paul's experience. It makes a big difference phenomenologically whether one's home intersection is leafy or bare; whether one is kissed on the cheek in a tree or on a swing; whether there is a perceptible world outside the SUV against which one can compare one's vision; whether the yards in one's neighborhood have sculpted plants, or fountains and tennis courts (the latter two would make the neighborhood straightforwardly upper-class); and whether there is a tree whose anthropomorphic characteristics can offer something to aspire to in marriage. And it also makes a big difference whether these instances can be linked only conceptually and symbolically, through the use of the tree imagery, or whether they can be linked more physically and literally, through mycorrhizal networks—whether Paul is dealing with five objects, or five facets of the same hyperobject. Through the trees that are its zones of aesthetic causality, the mycorrhizal hyperobject can offer shade to pedestrians, provide privacy to children, supply a contrasting view to car passengers, gesture toward residents' social class, and lend itself to setting a moral ideal for newlyweds.

It is important to note that the mycorrhizal hyperobject is not another metonymy for Nature, the pure Outside. Nature so conceived is defined in opposition to human society, as everything that human society is not. It has no positive definition. In contrast, the mycorrhizal hyperobject within the storyworld of *Taipei*, as I have described it above, is almost exclusively characterized constructively. It is no reservoir of conceptual Otherness, but an object with perceptible and imperceptible sides, whose facets can tune other objects through its zones of aesthetic causality to form new objects, each with their own organization of parts.

We have seen what can happen when objects attune to hyperobjects; and hyperobjects attuning to hyperobjects would look just like objects attuning to objects, for they are of the same scale; but what would it look like for an object to be caught between the attunements of two different hyperobjects?

THE CALL OF THE HYPEROBJECTS

When an object is part of a hyperobject, it still maintains in its imperceptible side facets that are unrelated to the hyperobject's zones of aesthetic causality and are thus unpresented to the hyperobject; this makes it susceptible to tuning by other (hyper)objects through their own zones. Moreover, since each object is a unity, it is not subsumed by the tuning hyperobject, and so its attunement is not exclusive—it can attune to multiple (hyper)objects at the same time. The fact that a bottle of vinegar is both an ingredient and a cleaner does not rend it in half. Finally, since a (hyper)object is but an organization of parts, the same parts could constitute two different (hyper)objects. I am part of academia, and if we view academia as a kind of society, with its own conventional wisdom and its own unspoken rules guiding interactions with strangers, we can say that the same parts that constitute academia also constitute a society, and that I am simultaneously part of both.

An object usually attunes to multiple (hyper)objects with no contradiction. But when a contradiction does arise, it is likelier to arise between hyperobject attunements, for hyperobjects, being so large, tune objects through a greater number of zones of aesthetic causality. I cannot at the same time be an inhabitant of both Earth and Proxima centauri b, the closest known exoplanet that might be habitable; even if dual residency could be granted on paper, the vast distance between them and my own mortality, given current technologies, would mean that I could never literally set foot on both in my own lifetime. For some more down-to-earth examples, we can turn to the work of Alain Badiou, a precursor of object-oriented ontology.[20] Badiou argues that there are four "truths," pursuing which humans can be "committed to a trans-specific procedure" that leads to "an experience of the inhuman": science, art, politics, and love.[21] The truth of each of these realms is arrived at by the procedure of "forcing": "That a term of the situation [to come] *forces* a statement of the subject-language means that the veracity of this statement in the situation to come is equivalent to the belonging of this term to the indiscernible part [of the original situation] which results from the generic procedure."[22] Translated into object-oriented terms: That a zone of aesthetic causality (term) of a new hyperobject (situation to come) tunes (forces) a facet (statement) of an object (subject-language) means that the location (veracity) of this facet of the object in the new hyperobject is equivalent to linking (belonging) this zone of aesthetic causality to that new hyperobject whose previously unpresented organization (indiscernible part) of the parts of the old hyperobject (the original situation) has now been presented (results) by the attunement (truth-procedure). In simple terms, one cannot simultaneously believe in two paradigms to explain the same experimental result, see the same value in two artistic formalisms, pursue two political ideals in the same society, or love two people in the exact same way and without changing oneself. The two paradigms, formalisms, ideals, and people—these zones of aesthetic causality of the respective hyperobjects of science, art, politics, and love—each tune us, we human objects; but the zones themselves are reorganized within a new hyperobject. Through the same zones (which, again, are themselves objects), we find ourselves attuning to new hyperobjects, differing from the old ones mainly in the organization of parts. This is to say that we are always already attuned to a hyperobject (the old hyperobject) when a new one comes along to tune us. And it is also to say that the parts of a new hyperobject are often already there, awaiting the right reorganization. A new hyperobject can be there before it is there; it is "a haunting, uncanny, *spectral* dimension,"[23] one that, as Derrida says, "begins before it begins": a "hauntology."[24] In *Taipei*, the hyperobject of the digital is haunted by the mycorrhizal hyperobject—a return to primordial roots. One cannot be both dualistic (the digital binary) and rhizomatic (the mycorrhizal network) at the same time,[25] and it remains to be seen how the old hyperobject can be reorganized into the new, how the haunted can become the haunter.

SUMMONING THE GHOST

The question of ethnic identity in the novel can shed light on this. For most of the novel, Paul's Taiwanese American identity is not a significant factor, mentioned only in passing. The most salient episode of identity formation in the novel is when, as an outgoing adolescent, he is relentlessly taunted for trying to be cool, which leads him to withdraw into himself so severely that he refuses to talk for several years.[26] In Frank Guan's reading, this transformation is only possible because of Paul's background as a second-generation Asian immigrant. Guan argues that the identity of a 1990s child in the United States had three determining factors: parents, school, and television.[27] If Paul, who (assuming the novel is set in or just before its year of publication) grew up in the 1990s, could not find a socially acceptable model of identity in his unassimilated parents, or in the overwhelmingly White (and sometimes Black) people on television, then being rejected, and not just rejected but "erased," by the school peers whose opinion he values most would quite predictably result in his complete withdrawal from society, only reemerging when the fourth factor of internet culture develops, the factor that ultimately leads him to think according to the logic of digital media. The digital hyperobject tunes Paul's identity through the zone of aesthetic causality of social acceptance, which is the very zone that was reorganized from the social hyperobject.

The same situation, I argue, applies to the two hyperobjects in the novel under discussion here: the digital and mycorrhizal hyperobjects. Reorganizing one into the other would open up an ecological facet of the storyworld that has hitherto been imperceptible. The zone of aesthetic causality to be shifted here would be the logic that guides Paul's thinking, from dualistic to rhizomatic. The climactic mushroom trip gestures toward how this might be done. After taking mushrooms with a higher dose of psilocybin than he expects, Paul begins to think that he has died, and his process of returning to life involves abandoning the analog for the digital. One of his last analog perceptions is of tap water, "cold, grasping, meticulous, aware," but already "he had no concept of what water felt like until he touched it"; and when he leaves the bathroom, he sees Erin's "movement as incremental—not continuous, but in frames per second." By the end of the episode, in the last line of the novel, he has been completely reborn as part of the digital hyperobject, and he says that "he felt 'grateful to be alive.'"[28] But the digital hyperobject has not entirely subsumed him: That last quotation, the celebration of life, is put in quotation marks, which in this novel can function either as quotes or as markers of irony. The possibility that this quote is ironic, that it is Paul's commenting on his own affect of celebration, is the facet of his thinking that has not been tuned by the digital hyperobject. And though the logic of his previous, analog, hyperobject is not evidently rhizomatic, perhaps the mycorrhizal hyperobject can tune him yet.

During his mushroom trip, as he strives to return to life, Paul thinks that he can do so "by discerning some code or pattern of connections in his memory, or remembering some of what had happened with a degree of chronology sufficient to re-enter the shape of his life, or sustaining a certain variety of memories in his consciousness long enough."[29] The code and, failing that, chronology of memories is dualistic: Either they fit entirely or they do not. Failing the chronology, Paul hopes for at least the correct variety of memories, an additive multiple state that Gilles Deleuze and Félix Guattari deride as "all the more total for being fragmented"[30]—also non-rhizomatic. But here's the rub: The novel inserts a section break *without resolving the question of whether these strategies are sufficient.* In line with the ambiguity of the final sentence, we do not actually know whether Paul has successfully resurrected himself into the digital hyperobject. I aver that the brief glimpses of the mycorrhizal hyperobject in the novel suggest, hauntingly, that he comes back alive via a third way: without ascertaining the "correct" dualistic or multiple organization. It is perhaps possible for Paul to have returned to life by attuning to the rhizomatic non-organization of the mycorrhizal hyperobject, a subtractive organization in which what is subtracted is organization itself—refusing to organize is itself a type of organizing. Paul can let his memories fall where they may. In the final sentence, what is on the other side of the irony of digital life is precisely mycorrhizal life, which is not its opposite but its reorganization. To reiterate a previous point, this would not be a pure Outside, especially not a pure Outside of thought, for the mycorrhizal hyperobject is not a lack defined against the digital and the analog—it is already there, in the very trees that are its zones of aesthetic causality, through which it has at multiple times tuned Paul.

We have previously said that the mycorrhizal hyperobject haunts the novel; but the above analysis has shown it to be not a spectral Other but an alternative organization. The logic of hyperobject reorganization is not hauntological but perspectivist. This bodes well for our efforts to mitigate global warming from within a neoliberal framework. After all, both global warming and neoliberalism are hyperobjects, whose massive scale have posed significant barriers to resolution through more conventional means. Morton says that "[t]o encounter an ecological entity is to be *haunted*,"[31] but perhaps it would be more productive to say that it is to encounter a radical (from Latin *radix*, "root"; cf. rhizome, from Ancient Greek *ῥίζωμα*, "mass of roots") reorganization. The extended—seemingly interminable—crisis of global warming, then, asks us to attune to this ecological hyperobject, to shift perspectives on the zone of aesthetic causality that is our mode of collectivity, so that it might be reorganized from the neoliberal hyperobject to the ecological one. The necessary parts have always already been there. We have begun before we have begun. The specter haunting *Taipei* is, for us, a hyperobject whose call we ought to heed.

NOTES

1. Chingshun J. Sheu, "What We Talk about When We Talk about New Media: Digital Subjectivity and Tao Lin's *Taipei*," *Textual Practice* 34, no. 8 (2020): 1269–1284, https://doi.org/10.1080/0950236X.2019.1580763/. Chingshun J. Sheu, *Hyperobject Reading, Scale Variance, and American Fiction in the Anthropocene* (Cham: Palgrave Macmillan, 2023), 71–104.

2. Timothy Clark, *Ecocriticism on the Edge: The Anthropocene as a Threshold Concept* (London: Bloomsbury, 2015), 64. Timothy Morton, *Ecology without Nature: Rethinking Environmental Aesthetics* (Cambridge, MA: Harvard University Press, 2007), 17, 79.

3. Tao Lin, *Taipei: A Novel* (New York: Vintage Books, 2013), 38.

4. Lin, *Taipei*, 38.

5. Lin, *Taipei*, 98.

6. Lin, *Taipei*, 103.

7. Lin, *Taipei*, 151.

8. Merlin Sheldrake, *Entangled Life: How Fungi Make Our Worlds, Change Our Minds, and Shape Our Futures* (New York: Random House, 2020), 149–174. Robert Macfarlane, *Underland: A Deep Time Journey* (New York: W. W. Norton and Company, 2019), 97–101.

9. Andrew Adamatzky, "Language of Fungi Derived from Their Electrical Spiking Activity," *Royal Society Open Science* 9, no. 4 (2022): 211926, https://doi.org/10.1098/rsos.211926.

10. Graham Harman, *Object-Oriented Ontology: A New Theory of Everything* (London: Pelican Books, 2017), 54.

11. Ian Bogost, quoted in Levi R. Bryant, *The Democracy of Objects* (London: Open Humanities Press, 2011), 19.

12. Jean-Paul Sartre, *Being and Nothingness: An Essay in Phenomenological Ontology*, trans. Sarah Richmond (London: Routledge, 2018), 7.

13. Timothy Morton, *The Ecological Thought* (Cambridge, MA: Harvard University Press, 2010), 41.

14. Graham Harman, *The Quadruple Object* (Alresford, UK: Zero Books, 2011), 114–115. Peter Wolfendale, *Object-Oriented Philosophy: The Noumenon's New Clothes* (Falmouth, UK: Urbanomic, 2014), 19.

15. Sheu, *Hyperobject Reading*, 5–13.

16. Timothy Morton, *Hyperobjects: Philosophy and Ecology after the End of the World* (Minneapolis: University of Minnesota Press, 2013), 171.

17. Graham Harman, *Skirmishes: With Friends, Enemies, and Neutrals* (Santa Barbara, CA: Punctum Books, 2020), 78.

18. Morton, *Hyperobjects*, 1.

19. Bryant, *Democracy of Objects*, 216.

20. Bryant, *Democracy of Objects*, 214; Alexander Galloway, "The Poverty of Philosophy: Realism and Post-Fordism," *Critical Inquiry* 39, no. 2 (2013): 348–349, https://doi.org/10.1086/668529/.

21. Alain Badiou, *Logics of Worlds: Being and Event II*, trans. Alberto Toscano (London: Continuum Books, 2009), 71.

22. Alain Badiou, *Being and Event*, trans. Oliver Feltham (London: Continuum Books, 2005), 403.

23. Timothy Morton, *Humankind: Solidarity with Nonhuman People* (London: Verso Books, 2017), 38.

24. Jacques Derrida, *Specters of Marx: The State of the Debt, the Work of Mourning and the New International*, trans. Peggy Kamuf (London: Routledge, 2006), 202.

25. Gilles Deleuze and Félix Guattari, *A Thousand Plateaus: Capitalism and Schizophrenia*, trans. Brian Massumi (Minneapolis: University of Minnesota Press, 1987), 5–7.

26. Lin, *Taipei*, 38–43.

27. Frank Guan, "Nobody's Protest Novel: On Tao Lin," *n+1*, no. 20 (2014), https://nplusonemag.com/issue-20/reviews/nobodys-protest-novel/. Last section.

28. Lin, *Taipei*, 247–248.

29. Lin, *Taipei*, 246.

30. Deleuze and Guattari, *Thousand Plateaus*, 6.

31. Morton, *Humankind*, 64.

BIBLIOGRAPHY

Adamatzky, Andrew. "Language of Fungi Derived from Their Electrical Spiking Activity." *Royal Society Open Science* 9, no. 4 (April 2022): 211926. https://doi.org/10.1098/rsos.211926.

Badiou, Alain. *Being and Event*. Translated by Oliver Feltham. London: Continuum Books, 2005.

Badiou, Alain. *Logics of Worlds: Being and Event II*. 2006. Translated by Alberto Toscano. London: Continuum Books, 2009.

Bryant, Levi R. *The Democracy of Objects*. London: Open Humanities Press, 2011.

Clark, Timothy. *Ecocriticism on the Edge: The Anthropocene as a Threshold Concept*. London: Bloomsbury, 2015.

Deleuze, Gilles, and Félix Guattari. *A Thousand Plateaus: Capitalism and Schizophrenia*. 1980. Translated by Brian Massumi. Minneapolis: University of Minnesota Press, 1987.

Derrida, Jacques. *Specters of Marx: The State of the Debt, the Work of Mourning and the New International*. Translated by Peggy Kamuf. London: Routledge, 2006.

Galloway, Alexander. "The Poverty of Philosophy: Realism and Post-Fordism." *Critical Inquiry* 39, no. 2 (winter 2013): 347–366. https://doi.org/10.1086/668529/.

Guan, Frank. "Nobody's Protest Novel: On Tao Lin." *n+1*, no. 20 (fall 2014). https://nplusonemag.com/issue-20/reviews/nobodys-protest-novel/.

Harman, Graham. *The Quadruple Object*. Alresford, UK: Zero Books, 2011.

Harman, Graham. *Object-Oriented Ontology: A New Theory of Everything*. London: Pelican Books, 2017.

Harman, Graham. *Skirmishes: With Friends, Enemies, and Neutrals*. Santa Barbara, CA: Punctum Books, 2020.

Lin, Tao. *Taipei: A Novel*. New York: Vintage Books, 2013.

Macfarlane, Robert. *Underland: A Deep Time Journey*. New York: W. W. Norton and Company, 2019.

Morton, Timothy. *Ecology without Nature: Rethinking Environmental Aesthetics*. Cambridge, MA: Harvard University Press, 2007.

Morton, Timothy. *The Ecological Thought*. Cambridge, MA: Harvard University Press, 2010.

Morton, Timothy. *Hyperobjects: Philosophy and Ecology after the End of the World*. Minneapolis: University of Minnesota Press, 2013.

Morton, Timothy. *Humankind: Solidarity with Nonhuman People*. London: Verso Books, 2017.

Sartre, Jean-Paul. *Being and Nothingness: An Essay in Phenomenological Ontology*. 1943. Translated by Sarah Richmond. London: Routledge, 2018.

Sheldrake, Merlin. *Entangled Life: How Fungi Make Our Worlds, Change Our Minds, and Shape Our Futures*. New York: Random House, 2020.

Sheu, Chingshun J. "What We Talk about When We Talk about New Media: Digital Subjectivity and Tao Lin's *Taipei*." *Textual Practice* 34, no. 8 (2020): 1269–1284, https://doi.org/10.1080/0950236X.2019.1580763/.

Sheu, Chingshun J. *Hyperobject Reading, Scale Variance, and American Fiction in the Anthropocene*. Cham: Palgrave Macmillan, 2023.

Wolfendale, Peter. *Object-Oriented Philosophy: The Noumenon's New Clothes*. Falmouth, UK: Urbanomic, 2014.

Index

About the Contributors

Kathryn Yalan Chang is Professor of English at National Taitung University, Taiwan. Her current research interests include critical plant studies, food and garden studies, and ecofeminism, wilderness healing, medical environmental humanities, material ecocriticism, environmental justice and activism, and critical animal studies. Her latest publications include "'Slowness' in the Anthropocene: Ecological Medicine in *Refuge and God's Hotel*," *Neohelicon* (2017), "'A Perfect Place to Die': Eco-Gothic, Darkness, and the Dark Sweet in Gus Van Sant's *The Sea of Trees*," *SARE: Southeast Asian Review of English* (2022), and "'Reframing Care' in the Age of a Novel Corona Virus: Food as Medicine on the Farms of Two Physician-Farmers," *The Bloomsbury Handbook to the Medical-Environmental Humanities* (2022), and "'Not Unloved Ones': Moss, Gender, and the Plantationocene in Elizabeth Gilbert's *The Signature of All Things*," *Review of English and American Literature* (2021).

Ysanne Chen is an MBA and MSc graduate student in the field of International Business at National Sun Yat-sen University in Taiwan and Johannes Kepler University Linz in Austria. She received Honorary Membership for Academic Excellence from the Phi Tau Phi Scholastic Honor Society in Taiwan (2022), as well as an Erasmus+ Scholarship while studying in Europe (2022). Her research interests center around cross-cultural consumer perception and often spans interdisciplinarily to studies of the humanities. For her master's thesis, she published quantitative research relating to global cultural authenticity perceptions of contemporary art and the regional cultural marketing representations of multicultural individuals. While working on an Indigenous literature project under Dr. Hsinya Huang, she also co-edited *Pondering the Pacific* (2022) and hosted an international seminar. In her former role as a research assistant, she has additionally worked at the English and American Literature Association in Taiwan from 2020 to 2022. She also previously directed the Foreign Languages and Literature Department's (NSYSU) annual musical theater production, sponsored by the American Institute in Taiwan, in 2017 and 2018.

Li-hsin Hsu is Professor of English at National Chengchi University, Taiwan. Her research interests include Emily Dickinson studies, Romanticism, Ecocriticism, and Modern Taiwan Poetry. She has published in a number of international journals, such as the *Emily Dickinson Journal*, *Romanticism*, *Studies in American Fiction*, *Ex-position*, and *Concentric*, and guest-edited a number of journal issues, including a special issue on Ecogothic Asia for *SARE: Southeast Asian Review of English* (July 2022), and a special issue on Asian Gothic for *The Wenshan Review* (June 2023). She has also contributed to a number of edited volumes, such as *Ephemeral Spectacles, Exhibition Spaces and Museums: 1750–1918* (2021) and *Romantic Environmental Sensibility: Nature, Class, Empire* (2022), on topics related to space, environment, and race. She is serving on the Emily Dickinson International Society (EDIS) board and will host the EDIS international conference in Taiwan in 2025 on Dickinson and Ecologies. She is also currently the organizer of the "Taiwan in the Anthropocene II" research group, sponsored by the National Science and Technology Council in Taiwan.

Hsinya Huang is Distinguished Professor of American and Comparative Literature, National Sun Yat-Sen University (NSYSU), Taiwan. She is former Dean of Arts and Humanities and Provost of Academic Affairs and Faculty Advancement, NSYSU, and served as Director General of International Cooperation and Science Education, Ministry of Science and Technology, Taiwan, 2018–2019. She is the author or editor of books and articles on Native American and Indigenous literatures, environmental humanities, and transnational studies, in Chinese or English, including *(De)Colonizing the Body: Disease, Empire*, and *(Alter)Native Medicine in Contemporary Native American Women's Writings* (2004) and *Native North American Literatures: Reflections on Multiculturalism* (2009), *Aspects of Transnational and Indigenous Cultures* (2014), *Ocean and Ecology in the Trans-Pacific Context* (2016), *Chinese Railroad Workers: Recovery and Representation* (2017), *Diaspora, Memory, and Resurgence: Trans-Pacific Indigenous Writing and Practice* (2022), *Pacific Literature as World Literature* (2023), and *Radiation Ecologies in Trans-Pacific Indigenous Texts: After Hiroshima* (2024, forthcoming). She serves on the Advisory Board of *Verge: Studies in Global Asias*, *The Journal of Transnational American Studies* and *Routledge series on Transnational Indigenous Perspectives*, and on the Editorial Board of *Transmotion: A Journal of Vizenorian Indigenous Studies* and is Director of Asia Pacific Observatory of Humanities for the Environment Global Network.

Rose Hsiu-li Juan is a Professor Emerita in the Department of Foreign Languages and Literatures, at National Chung Hsing University, Taichung, Taiwan. Her academic service includes board member of the English and American Literature Association in the Republic of China and board member of The Association for the Study of Literature and Environment in ROC (ASLE-Taiwan). She was the Deputy Dean of the College of Liberal Arts (NCHU), Director of the Humanities Center (NCHU), and chairperson of DFLL. Her research interests include Indigenous literature and culture, ecocriticism, and transnational Taiwan literature. Her publications appear in journals in Taiwan and abroad and as book chapters: *Chung-Wai Literary Quarterly*, *Review of English and American Literature*, *Animal Dreams in Aesthetic Media*, *Ecocriticism in Taiwan: Identity, Environment, and the Arts* (Lexington Books), *Typologizing the Dream*, *Writing the Dream*, volumes of selected papers of ICLA Congress, *Mapping Native North American Literatures: Reflections on Multiculturalism* (Taiwan, co-editor with Hsinya Huang), and *An Introduction to Ecoliterature* (Taiwan). Her creative writings won literary prizes and appeared in prose selections of the year.

Pei-Wen Clio Kao is an Associate Professor at the Department of Foreign Languages and Literature of National Ilan University, Taiwan (R.O.C.). Her research specialty lies in Conrad Studies, Faulkner Studies, and Literary Modernism. She completed her PhD from National Chengchi University, Taiwan, in 2015. She did overseas studies at Connecticut University (US) as an exchange PhD student and was a visiting scholar at the Center for Faulkner Studies at Southeast Missouri State University (US). She has published articles on Modernist writers, Conrad and Faulkner in international journals and in international book series, including *Conradiana*, *The Conradian*, *CLCWeb*, *VTU Review*, *Review of English and American Literature (REAL)*, *English Literature: Theories, Interpretations, Contexts*, and *Joseph Conrad Today*. She is the winner of the 2024 Eudora Welty Review award.

Iping Liang is Professor of English and American Studies in the Department of English at National Taiwan Normal University. She specializes in the areas of Asian American literature, Native American literature, Critical Plant Studies, and Archipelagic American Studies. As the former president of the Association for the Study of Literature and Environment in Taiwan (ASLE-Taiwan), she organized the 2020–2023 Critical Plant Studies (CPS) research group and thus launched CPS in Taiwan. In addition, she is the author/editor of numerous monographs, including *Diving into Nature: Literature and Ecocriticism in the Anthropocene* (2023), *Mushroom Clouds: Militarization and Ecocritical Approaches to the Environment in East Asia* (2021), *I'm Migrant: New Perspective of Overseas Chinese Studies* (2018), *Asia/*

Americas: Asian American Literatures in Taiwan (2013), and *Ghost Dances: Toward a Native American Gothic* (2006). Her current research project investigates plant narratives in Pacific Asia in the era of the Cold War.

Chia-hua Lin is a PhD candidate in the English Department of the University of Hawai'i at Mānoa. She is the recipient of the 2018 Fulbright Graduate Study Grant, the 2020 Government Scholarship to Study Abroad (GSSA) from the Taiwanese Ministry of Education, and the 2023 Taiwanese Overseas Pioneers (TOP) Grants from the National Science and Technology Council. Her research interests include Trans-Indigenous Studies, Ecopoetics, Hawaiian Literature, and Pacific Literature. She currently works as the secretary for the Asia Pacific Observatory of *Humanities for the Environment*. She is the co-editor of *Chinese Railroad Workers in North America: Recovery and Representation* (2017), and *Pacific Literature as World Literature* (2023).

Yih-Ren Lin is a professor at the Graduate Institute of Museum Studies, Taipei National University of the Arts. Trained both as an ecologist and geographer at University College London, UK, his research fields include social nature, cultural geography, political ecology, action research methodology, traditional ecological knowledge, and ecological medicine. He had served as the director of the Graduate Institute of Humanities in Medicine at Taipei Medical University and taught there for the past ten years. Before that, he taught at the Department of Ecological Humanities of Providence University and established the General Education Centre and Austronesian Research Centre. Over the past two decades, he has been actively engaged in numerous Indigenous community development projects and is currently working with the *Millet Ark Initiative* to promote the resilient livelihood of agricultural and natural resources management with Tayal people of Taiwan. He is the filmmaker of the Houston Film Festival Remi Awards–winning film *Once Upon a Time* in 2009. Besides academic writing, he writes popular column articles for magazines and social media in terms of policy critique for the land rights issues of Indigenous peoples and the environment.

Li-Ru Lu is a professor of English at National Sun Yat-sen University, Taiwan. Her research interests include nineteenth-century Taiwan studies and environmental humanities, English and American nature writing, energy humanities, and early American literature. She is the author of a book titled *Uncovering New Ground for American Nature Writing: Discourses of Natural History from John Bartram to Wilson Flagg* (2014). Her other publications include research articles in peer-reviewed journals in Taiwan, the United States, and Malaysia. In 2018 and 2022, her research articles were respectively published in *SARE: Southeast Asian Review of English*;

her 2018 article is titled "Travel, Coal, and Imperialism: Delineations of Formosan Coal in Nineteenth-Century English Travelers' Coal Texts"; her 2022 article is entitled "Life Writing of Marine Animals in Cuthbert Collingwood." She is also the author of book chapters in a number of edited volumes, such as *Writing the Environment in Nineteenth-Century American Literature: The Ecological Awareness of Early Scribes of Nature* (2015), *South East Asian Ecocriticism: Theories, Practices, Prospects* (2018), *Life Mapping as Cultural Legacy* (2021), *Critical Plant Studies in Taiwan* (2024), and *Posthuman Southeast Asia* (2024 or 2025, forthcoming).

Pagung Tomi is a self-sufficient Indigenous researcher of Tayal people in Taiwan. Over the past thirty years, she has devoted herself to the cultural revitalization work of Tayal people including inheriting the ancient singing *Lmuhuw*, traditional territory mapping, seeds collection of traditional crops and related practices, facilitation of co-op for local communities, and so forth. She is the main figure of the *Millet Ark Initiative* aiming to revitalize Tayal people's millet culture and traditional natural farming knowledge. She co-authored a millet culture academic paper *Situating Indigenous Resilience: Climate Change and Tayal's "Millet Ark" Action in Taiwan* in 2020. She is now a postgraduate student in the Department of Ethnology, National Chengchi University.

Theodoor A. M. Richard is Research Fellow at National Chung Hsing University, Taichung, Taiwan, with a PhD in Taiwan and Transcultural Studies. His dissertation was a longitudinal autoethnographic study of experiencing transculturality when moving from The Netherlands to Taiwan. Dr. Richard's research focuses on how culture forms in glocalized times as a function of (borderless) mobility in the Anthropocene, and how sustainability thinking is to shape our cultural formation in order to avoid the climate change crisis from an interdisciplinary perspective. Special topics of interest are sustainability culture, new climate ethics, agroecological transition, regenerative agriculture, transculturality and transcultural system thinking, diasporic culture, borderlessness, and how belief systems decide the way we act. Recent publications include: "Redefining Taiwan New Cinema as Borderless Transcultural Cinema From Taiwan" in *Tamkang Review* (2022), and "On the Ethical Challenge of the Predicament of AI and the Perspectives Offered by Buddhism" in *Exchanges*, an Interdisciplinary Journal at the University of Warwick (2022).

Stephen Roddy is a Professor of Languages, Literatures, and Cultures at the University of San Francisco. His research has focused on intersections between institutional change and literary expression in late pre-modern China and Japan. Recent publications include a translation (with Ying Wang) of a 1651 *chuanqi* drama on the theme of female same-sex love, *The Fragrant Companions* (*Lianxiangban*), and studies of poetry and prose essays by Ji Yun (1724–1805), Gong Zizhen (1793–1841), and Yu Yue (1821–1906).

Chingshun J. Sheu is an Assistant Professor of Applied English at Ming Chuan University, specializing in existentialism, narratology, and contemporary American fiction. He has published on William Gaddis, Joshua Ferris, Tao Lin, Orson Scott Card, Marilynne Robinson, and Taiwanese author Chang Hsiu-ya. His book, *Hyperobject Reading, Scale Variance, and American Fiction in the Anthropocene*, was published by Palgrave Macmillan in 2023.

Weibon Wu is an Associate Professor in the Department of Applied English at National Quemoy University. He holds a PhD in English Literature from Tamkang University. He was also a research scholar at the University of Nevada, Reno, where he studied contemporary American and environmental literature. He has published essays in *Tamkang Studies of Foreign Languages and Literatures*, *Shih Shin Journal of Humanities,* and so on. His current research interests include animation and sci-fi novels, young adult literature, and warzone afforestation in Quemoy (Jinmen). He is also the author of a chapter titled "Blue Blood: A 'Silent War' at the Edge of the Sea" in the book *Mushroom Clouds: Ecocritical Approaches to Militarization and the Environment in East Asia* (2021), and an essay titled "Unsettling the Yoke of Humanity: T. C. Boyle's Strange Man, Apes, and Dog Woman" in *The Wenshan Review of Literature and Culture* (2018).

Ya-feng Wu has been a Professor of English, in the Department of Foreign Languages and Literatures, National Taiwan University, since 2012. She obtained her PhD from Glasgow University, United Kingdom. Having specialized in Romantic Literature and Art and Gothic Revival Architecture, her current projects include Natural History, East and West Encounter, and Literature and Environment. Her publications appear in *The Wordsworth Circle*, *Chung-Wai Literary Quarterly*, *Comparative Literature and Culture*, *Ex-position*, and so on.

www.ingramcontent.com/pod-product-compliance
Ingram Content Group UK Ltd.
Pitfield, Milton Keynes, MK11 3LW, UK
UKHW041452071026
15587UKWH00010B/816

9 781666 935363